FORCE OF HABIT

FORCE OF HABIT

Unleash Your Power by Developing Great Habits

TAMSIN ASTOR, PHD

CORAL GABLES

Published by Mango Publishing Group, a division of Mango Media Inc.

Cover and Layout Design: Elina Diaz

For permission requests, please contact the publisher at:

Mango Publishing Group
2850 Douglas Road, 3rd Floor
Coral Gables, FL 33134 USA
info@mango.bz

For special orders, quantity sales, course adoptions and corporate sales, please email the publisher at sales@mango.bz. For trade and wholesale sales, please contact Ingram Publisher Services at customer.service@ingramcontent.com or +1.800.509.4887.

Force of Habit: Unleash Your Power by Developing Great Habits

Library of Congress Cataloging

ISBN: (print) 978-1-63353-786-6 (ebook) 978-1-63353-787-3
Library of Congress Control Number: 2018946829
BISAC category code: SEL044000 — SELF-HELP / Self-Management / General

Printed in the United States of America

CONTENTS

PREFACE

This book started in the fall of 2016, as an expression of my personal journey and what I considered to be the best tools to enact the key habits—*sleeping, eating, exercising*, and *working on your relationship with yourself and others*—to manage your life and navigate it to the best of your ability.

As this book details, I am no stranger to heartbreak and difficulties—a child with cancer, the death of someone I love, my marriage ending. Each of these events shattered the cage of my life and each time I chose to rebuild and start again. By summer 2017, I thought I was free and living in a strong, unbreakable cage.

But then, during the fall of 2017, my cage was shattered again.

I went through a number of truly upsetting experiences. These involved lots of deep personal boundary-breaking interactions with people in my life which fit in with the cultural tide of the #metoo movement and culminated with a debilitating injury which forced me to lie on my back for almost a month.

And this time, like every time before, when my cage was shaken and I felt the ground give way between my feet, I have gotten back up again.

How?

By developing and practicing the habits described in this book.

These habits created a massive supportive shift in my personal self-care and pushed me to face myself and how I live with myself, treat myself and share myself with people in my life. I am no longer the last person I notice or nourish. I have learned how to unleash my power.

The tools to create the healthy habits that I offer in this book work because they are my unique combination of the extensive experience I have in both the eastern and western ways of studying, managing, and creating sustainable patterns in our mind-body systems.

Each time my cage shattered and I thought, “It can’t get any worse than this,” I turned to these habits and found greater and greater strength, clarity, and grace in how to engage with my life.

The word “force” compels me to remind you that it means to break open, that it is strength and energy, might and effort. That is what will happen to you as you start to face yourself and your habits. You will see who you are and how you show up for yourself, which is often markedly different from how we show up for others.

When you start to enact these healthy habits, you will unleash your unique power. Power for most of us means being super successful at whatever we are focused on—our work, our parenting, our relationships, our role as citizens.

“You will never change your life until you change something you do daily. The secret of your success is found in your daily routine.”

—Darren Hardy

When you develop great habits, you establish regular tendencies or practices. I am your Chief Habit Scientist, guiding you to cultivate, reflect, practice, and ultimately become your own habit scientist.

This book details the main ways that you can unleash the power of your habits to break your own cage and create the life you want and the life you need, not the life that others expect from you.

INTRODUCTION

THE SHOULDS AND THE OUGHTS: WHY THEY DON'T SERVE YOU

I should walk the dog three times a day. I should file my nails. I should clean my fridge. I should learn to speak Italian. I should bake bread from scratch for my kids. I should read the sci-fi novel my cousin lent me (yes, Ian, that's you!). I should buy some cellulite cream. I should drink less coffee. I should use anti-wrinkle eye serum. I should get another certification. I should throw parties. I should stop drinking cheap wine. I should buy new underwear. I should vacuum up the dog hair more frequently. I should clean the gutters, fix the muffler on my car, trim the hedge, call my mother more, take a painting class, learn engineering, stop killing spiders, discontinue using white flour, and hire a housekeeper. I should, I should, I should…

I had an amazing childhood—love, happy parents, a good education, family and friends, travel, and intellectual and emotional learning. And I always had a very strong sense of what was right and wrong—not from a religious perspective, but rather from an innate sense of fairness (this sense is now bubbling up in my tween- and teenage kids, so I am learning to parent my younger self—hah!). This sense of what's right and wrong followed me through my life and was often an incapacitating shadow. I should weigh this much, I should look this way, I should want this kind of job. I should, I should, I should. But there was also a deep part of me that rebelled against the "shoulds."

> Say out loud three things that make you happy—not what "ought" to make you happy but what REALLY makes you happy! GO!

I first felt the suffocating grip of the "shoulds" as a teenager when I realized I didn't want to conform and be the kind of girl I was supposed to be according to my upbringing, class, and so on. So, at age fourteen, during my three weeks of summer school in Germany, I had two more holes pierced in my left earlobe. I dyed my hair many different shades and experimented with crazy outfits (my mother even asked me to walk apart from her down the street because I looked so kooky!). I got my first tattoo (illegally) at sixteen. Then I pierced my nose. Then I figured out a way to leave London and live in America when I was eighteen. I had my belly button pierced and went out dancing almost every night of the week (but still showed up to class at 8:00 a.m.!).

Close your eyes.
Breathe up (inhale) and down (exhale) your spine.

Ask yourself:
How do I feel?
Say it out loud:

"I feel ______."
Repeat twice more.
Then ask yourself:
Is there an action I can take that is related to how I feel? Do it.

In 1998, when I was finishing my undergraduate degree in psychology in London and most of my friends were interviewing for jobs in finance and advertising (the two most popular areas given the extensive math we did and our supposed ability to understand people and convince them to purchase—neither of which were appealing to my almost-twenty-two-year-old self), I was contemplating strategies to make my way to India. It was then that my lovely and excitable undergraduate project supervisor, a skinny, bicycle-riding vegetarian, suggested we have a chat about graduate school.

Over a bottle of red wine, which he had carefully warmed under the heat of his desk lamp, we constructed a plan to dive in more deeply to the "sexy" data (my professor's words!) that I had collected in my undergraduate project. I had the pleasure of sitting with him in the two comfy corner chairs, a floor lamp between us, as dusk fell and the mass of London's bright evening lights began to shine. In the fall, I started a PhD program.

Doing a PhD allowed me to balance the rebel within.

I spent my days reading, writing, gluing electrodes on people's heads, and listening to academics argue about the relative merits of their theories—all while wearing whatever I wanted, working wherever and at whatever time of day was appealing to me—while in the background, I knew I was on a "proper" career path to becoming an academic.

And then, life got busy fast. I was living with a fellow academic, and before I knew it, I was living in America with a husband and a PhD. And the rebel voice in me was anesthetized and the "shoulds" really took over.

Feeling blah and "should-upon?" Turn on your favorite piece of music and DANCE—go crazy!!

The stress built. By 2003, I was working full-time as a postdoctoral research fellow. I had two sons in two years (2004 and 2006). I had "proved" to the US Immigration and Naturalization Service that I was in a real marriage, not a green-card marriage, and I worked hard to be the best wife, mom, academic, and teacher that I could possibly be. The "shoulds" had really sunk their teeth in by then: I should make my baby organic food to take to his day care, I should teach my baby sign language so he can communicate with me, I should lose ten pounds, I should want this, I should do that…

After five years in Missouri, we found ourselves in Cleveland, Ohio. I had decided to leave academia. The politics, the fiefdoms, the battles for funding, the so-called peer-review game that meant getting published in the "right" kinds of journals had lost their appeal, and so I became the stay-at-home mom of two little energetic boys. But the stress and the "shoulds" kept getting stronger. I realized that stay-at-home motherhood was not my thing, so I trained as a yoga teacher and educator (using yoga and meditation as tools for teachers and counselors to manage classrooms and working with kids who had attention deficit hyperactivity disorder and autism spectrum disorder). I also started teaching people how to be yoga teachers.

As the stress grew and the "shoulds" got weightier, and I felt the need to control and manage, I essentially lost control: my younger son, who was two years old, was diagnosed with cancer. This shook my sense of "should" to the core—but more about that later.

The "shoulds" are your inner critic. Give them a name and a job that fits their type of critical nature. Maybe a copyeditor called Kai? Or a drill sergeant called Campbell? When you hear the voice of the inner critic, give it a job, so you can get on with your life and work!

WHAT DO I MEAN BY THE "SHOULDS"?

The "shoulds" are those feelings that can take over your thoughts—the inner critic. They tell you how you should look, feel, and behave. They tell you what you should want and what your life should be like. Often, these feelings are associated with societal norms or media-portrayed images.

As I worked to balance the "shoulds," I found myself doing more and more yoga. It was the first form of exercise I'd done that revealed to me the flow of mind and body. I started spending more time with yogis as I embarked on my yoga teacher training.

A yogi friend gave me some songs by the band Wookiefoot a few years ago. One of their songs is called "Don't Should on Me":

When you should on your friends it's bad for their health

But you got to be careful not to should on yourself...

Don't you should on me and I won't drop my shoulds on you

Shoulding on yourself can make you feel anxious! Cultivate equal breath: make the inhale and & exhale the same length by counting your breaths.

I loved this advice—not only should I not "should" on myself, I shouldn't "should" on my friends (or family)!

I continued to explore this concept more deeply while I did my Executive Coaching certification, where it showed up in a slightly different way: the Ought Self, the Actual Self, and the Ideal Self. These are personal standards, based on the self-discrepancy theory proposed by Edward Higgins in 1987, which help to motivate us and get on in life.[1]

Shoulding on other people doesn't usually go so well. How about finding three things that person does for you that you can be grateful about?!

I ought to look like this, make this much money, weigh this much, like this person. Whatever it was, the "shoulds" I placed on myself were *way* more prolific and stinky than the ones other people placed on me. As I dove into yoga, Buddhism, and meditation, my self-work really began. Letting go of my own ought self and finding ways to stop "shoulding" on myself became a driving force in my personal life. As I struggled to parent two little boys and then a daughter, all while navigating the difficulties of a profoundly unhappy spouse who was struggling professionally, I worked hard on myself so that I could take on each of the roles in my life with as much joy and energy as possible.

1 Higgins, E. Tory. "Self-discrepancy: A Theory Relating Self and Affect." *Psychological Review* 94, no. 3 (1987): 319–40.

So, this book, if you will, is a balance between nurturing your wild freedom and constructing limits. I learned that, by paying unflinching attention to my habits, my goals, my life—how I think, feel, and act—and by creating and committing to good and healthy habits, I could change my life. By erecting boundaries around and connections between particular behaviors, I've learned how to create a wonderful, joyful life.

Creating a happy, joyful life of ease is really about creating a framework that cultivates freedom. Releasing you from the "oughts" and "shoulds" is a good start. These do not serve you.

HASHTAGS #Shoulds #Oughts #OughtSelf #IdealSelf #DontShouldOnMe #DontShouldOnYourself #ActualSelf #Joy #Freedom

CHAPTER 1

MOTIVATION

WHY ARE YOU READING THIS BOOK AND WHY DO YOU CARE?

Perhaps you're just not feeling quite right. Like life is living you instead of you living it. You might feel like you're not fully present in all the roles you might be juggling—worker, partner, friend, parent, child, individual. You've got this sense of feeling overwhelmed and find it hard to both turn off at the end of the day and switch with ease between these roles. You see photos on Facebook or Instagram and wonder how people seem to have their shit together enough to have a family meal, to make time for yoga, to go on a date. You just KNOW that there's more to life. You sense that things could feel more easeful, more productive, more joyful.

If that's the case, then this book is for you.

Maybe you're reading this book because, like me, you've had a facedown-in-the-arena moment, as author Brené Brown calls it.[2] My facedown-in-the-arena moment started when my son was diagnosed with cancer. I put myself out there as a parent, I did everything I was supposed to do while I was pregnant—ate healthy food, exercised, slept well—and then I did everything I was supposed to do when he was a baby and young child. I breastfed him for twenty-one months, wore him in a sling on my body so he developed a healthy attachment to me, made organic food, laid him on his back to sleep, put up baby gates and strapped him into his five-point car harness, took him to baby yoga classes and music classes, avoided choking hazards, took him to swim lessons, rocked him, played with him, taught him how to engage with his big brother. You name it, I did it.

---

2 Brown, Brené. *Daring Greatly: How the Courage to Be Vulnerable Transforms the Way We Live, Love, Parent, and Lead.* New York, NY: Avery, 2015.

And then…

I fell facedown and lost control. I took time to look at the view from facedown. The world looks different when you're lying in the hospital with a two-year-old boy who has a nasogastric tube coming out of his nose and taped to his cheek, which he keeps tugging out and you and the doctors have to restrain him to re-insert it. The view is not pretty when your two-year-old has an epidural inserted in his spine, a wound halfway across his abdomen with a drainage line, and he keeps pulling out the IV lines in his arms and ankles. The view from facedown shifted and taught me more as his hair fell out and he'd wake with mouthfuls of his own hair saying, "Mama, there are spiders in my mouth." It shifted when he screamed with pain as he urinated the burning chemotherapy drugs out of his system, and when I held him as he writhed in pain when his intestines herniated two months after the initial cancer surgery.

> Look at Maslow's Hierarchy of Needs (page 22). Choose a level where you feel your needs are not being met and take a step towards changing that!!

When you lose control and you're facedown, there are some choices to be made. You can wallow and stay there, letting the shitty things that happen define your entire life. Or, you could wallow for a bit and then get up, walk away, and never look back or reflect, just box the experience into the past. Or, you could wallow for a moment, reflect on the view, agree to make some changes, and then get up.

We all know people who have fallen into those categories. Consider that divorcée who lives on your block who is still bitching and moaning about her ex-husband twenty-five years later (what you might call living through the rearview mirror). Then there are those people who, when they have a terrible or life-changing experience, box that experience into the past. They don't learn from it or reflect on how they could do things differently, and instead find themselves repeating the mistakes of the past, continuously living the same patterns.

Have you ever read Viktor Frankl's *Man's Search for Meaning*? This book is one of the most powerful works to contemplate the horrors of the World War II prison camps, to reflect on that awful experience, and then to contemplate how to use that knowledge for creating a meaningful life.

Frankl, who had trained in neurology and psychiatry with a focus on depression and suicide, wrote his book after surviving Auschwitz. In the book, he details a terribly moving passage in which he and another prisoner talk about their wives and how they hope that their experiences are not as awful as their own while they are walking to work, helping each other, and being beaten by the Auschwitz guards:

> "The truth—that love is the ultimate and the highest goal to which man can aspire… In a position of utter desolation, when man cannot express himself in positive action, when his only achievement may consist in enduring his sufferings in the right way—an honorable way—in such a position man can, through loving contemplation of the image he carries of his beloved, achieve fulfillment."[3]

Frankl's wife died, as did his parents and brother. But he directed his life's work to research and helping people understand the vital importance of finding meaning in every kind of existence—even the most appalling ones—and, therefore, to find reasons to continue living.

> List two "crimes against wisdom" (a.k.a. *prajnaparadha*, page 19) that you do, like drive-through doughnuts or scrolling through your ex's social media page at 11:00 p.m., and decide to do something that is empowering (perhaps opposite).

My experience, was of course, nowhere near the horror of Frankl's and I am not in any way trying to draw parallels. But my sad and difficult experiences led me to this realization: I wanted to understand so I could move forward with knowledge, which is how I unleash my power.

I started studying so I could understand different perspectives on life, death, mind, body, spirit, and soul. I meditated, took classes, talked to people, and read books. I started to find ways to manage my mind, and yoga helped my body start to feel more connected to itself and with my mind, but I still struggled to figure out why I didn't feel completely healthy. I needed to figure out what was going on.

As I slowly figured out the changes I needed to make and started instituting them in my life, I would fall off the wagon and fail to follow through with the new habits that I had been cultivating and that I knew would make my life better. When I started studying

---

3 Frankl, Viktor E. *Man's Search for Meaning*. Boston, MA: Beacon Press, 2006.

Ayurveda, the sister science of yoga that basically means the science (*veda*) of life (*ayus*), I discovered a great word in Sanskrit.

Prajnaparadha, pronounced "prudge-nah-purr-ahd-ha," means a crime against wisdom. I would commit this by staying up way too late, drinking way too much coffee late in the day, or trying to have a discussion with my then-husband about how our marriage was failing at 9:00 p.m. and then not sleeping well. Duh. The more I studied positive habits, and whether they were sticking for me or my clients, the more I realized that I had to get my head around the concept of motivation.

One of the first constructs I considered was that perhaps I had struggled because I was stuck in reliving my guilt from past "incorrect actions." This was definitely prevalent when my son was diagnosed with cancer—I found my meditation practice had a profound effect on my feelings about myself, my actions, and whether I had inadvertently caused the cancer by, I don't know, standing too close to the microwave when I was pregnant. Perhaps I needed to become a hedonist (someone who thinks the pursuit of pleasure is the most important thing in life) and start living purely for this moment and this moment alone! Or maybe a nihilist (someone who believes life is meaningless, so it doesn't matter what you do or how you act)? Hmm.

Philip Zimbardo, leader of the famous Stanford Prison Experiment—a study in which groups of undergraduate students were randomly assigned roles as prison guards or prisoners in an examination of perceived power versus innate traits—has done some interesting research on how we make decisions, which relates to this concept of motivation.[4] Specifically, in a 2009 TED Talk, he explores whether we are overly focusing on the future, the past, or the present when we decide what to do or not to do.[5]

If we are past-oriented, we are using our memories of similar decisions (either positive or negative); if we are present-oriented, we are focused on the action in front of us (either hedonistic or nihilistic); and if we are future-oriented, we are considering the costs and benefits for the future (either goal-oriented or transcendental, the idea that life begins after death). Zimbardo's research suggests

4 "The Stanford Prison Experiment: 40 Years Later." Stanford University Libraries. https://library.stanford.edu/spc/exhibitspublications/past-exhibits/stanford-prison-experiment-40-years-later.

5 Zimbardo, Philip. "The Psychology of Time." TED Talk. February 2009. https://www.ted.com/talks/philip_zimbardo_prescribes_a_healthy_take_on_time.

that we want this winning combination: past-positive, to give us roots; present-hedonistic, to give us pleasure and energy to explore; and future-goal-oriented, to give us the wings to soar!

THE STICK OR THE CARROT?

Research shows that we are often more motivated by the stick (pain) than the carrot (pleasure). For example, the pain of losing $100 is much stronger and more emotionally valent than the pleasure of winning $100. As a young kid, I tended to be more motivated by the stick than the carrot. I was fearful of breaking the rules because I did not want to deal with the consequences. I turned in my homework because the pain of doing it was less than the pain of my teacher asking me why I hadn't done it, in front of the class.

> Track your successes: use a big calendar and hang it somewhere to look at every day and then mark your weight loss, your meditation practices, your trips to the gym/yoga class. My favorite location is the bathroom—while I brush my teeth, I check this out!

I have found this to be a useful framework for making the initial shift in the direction of a new habit. For example, use it to set a goal for yourself, such as "I will make it to the gym X times this month," or "I will weigh X pounds by this date." Then, instead of a reward for making your goal happen, set up something that you don't want to happen (the stick). For example, give $100 to a charity which does exactly the opposite of what you value—that is, they have an ethical framework that is anathema to your way of engaging in the world. Write the check, post-date it and give it to a friend. Then get to making your goal happen. The thought of a negative consequence is often more motivating than rewarding yourself with a massage or a new pair of sunglasses (or whatever your positive reward might look like).

YOUR BIG WHY: THE PURUSHARTAS

This next concept feeds into why you're making this shift, and ties back into why you're reading this book in the first place.

This kind of motivation will invariably shift as your life changes, but the technique for getting at your big "why" is to look at the bigger context of your life and really get clarity around your vision for it. Where do you want to live, what do you want to do, and so on? For example, I had a client who wanted to lose weight, sleep better, and

Connect your healthy habits to your big why. E.g., I go to the gym at 5:30 a.m. because: I like to feel fit, I like to look good and feel good, I like to have energy for my job, my mother has osteoporosis and I'm not going to get it, I want to be able to chase my kids around the park!

have more energy so that she could spend the next twenty-five years of her life being fit and engaged as a grandmother. Her big "why" was about how sorting out her habits would create the kind of life that she wanted.

I do the work I do because I want to make changes in people's lives. Why do I do it this way (as a coach, author, and speaker), rather than as a therapist, doctor, or variety of other care-based professions that I am drawn to? Because my bigger "why" is freedom about where and when I work. I love to collect my kids from school, walk my dog during the day, and travel both with my kids and on my own. Doing this kind of work gives me the freedom not to be tied to a physical location or conventional nine-to-five hours.

A more formal framework for this can be provided by the yogic framework known as the *purushartas*—literally, the desires of the soul. Specifically, the purushartas can be considered the four aims of life. The **first aim is Dharma**: your life's purpose. The **second is Artha**: your worth and your desire for income and security. The **third is Kama**: desires, pleasures, and psychological needs (like Kama Sutra!). The **fourth is Moksha**: liberation, freedom, and spiritual enlightenment. As my friend, colleague, and fellow Yoga-Ayurveda teacher Marc Holzman notes, it's much easier to get up at 5:00 a.m. and meditate if you're clear on your *purushartas*!

ACTING FOR OTHERS

Put on your oxygen mask first: choose one daily habit that makes you feel amazing and commit to it. For example: wake before your kids and sit quietly and read a book, take a yoga class, or lie in the bath and listen to a podcast for twenty minutes.

When you take refuge as a Buddhist, you commit to saying the Bodhisattva vow, which basically states that you wish to continue being reborn until everyone is enlightened. Essentially, it is a repeated reminder to continue practicing, acting, and dedicating your practices and work to others. There are many people in the world and throughout history, such as Mother Theresa or Nelson Mandela, whose motivation for changing and connecting with the world was driven by service to others. Acting for others is another major driving factor—motivation—which can prove very powerful.

My work around this as the basis for my actions was that I put the needs of my former husband and my children before my own. I was not putting on my oxygen mask first, as it were. This led me to a place of massive depletion, where I was not meeting my own needs and so I struggled. Now I balance the needs of others with my own. I am fully cognizant of the fact that I need to fill my own cup first, if I am going to show up and be able to act for others from my full capacity. As one of my coach friends phrased it, fill your own cup so it overflows for those in your life. My children, friends, lovers, family, and clients are the recipients of my overflowing cup, and not the ones who get priority to drink from my cup first!

MASLOW'S HIERARCHY OF NEEDS

I got very good at reacting to and fulfilling the needs of those around me while ignoring my own needs. A large part of my self-work has been to really understand needs. I wanted to understand what my needs were and whether I was reacting to them appropriately.

Maslow's theory is a developmental model of needs and its original form consists of a pyramid with five layers: Physiological, Safety, Love/Belonging, Esteem, and Self-Actualization (to Self-Transcendence), from bottom to top.[6] This theory has been argued about and reconsidered and had other layers added to it. It is clear that we can develop and redefine more than one layer at a time; we are not confined to only one level and a progression from lower to higher levels. Maslow's theory has also been criticized for being ethnocentric (i.e., taking the view of only one ethnic group), which is completely valid, especially when you consider cultures and ethnic groups who place different values on different needs.

The reason I present Maslow's theory here is to make the point that, when you are making changes to habits in your life, which essentially requires you to shift and evolve who you are and how you show up in the world, it can be helpful to consider whether you are putting more or less attention on one of these needs. For me, as a super-nerd, I made the mistake of starting with the self-actualization and self-transcendence needs. I spent a lot of time thinking about my experiences, my life, and how I was showing up and engaging as a mother, wife, friend, and employee, but I failed, initially, to face, embrace, and ultimately make the necessary changes in the other categories of needs. So, I present this theory as something to

6 Maslow, A.H. "A Theory of Human Motivation." *Psychological Review* 50, no. 4. 370–96.

consider in terms of whether your attention and energy is focused too narrowly.

- **Physiological needs** are considered to be the most important and include air, food, water, and basic shelter. (After the basic physiological needs are met for the most part, the needs of safety become more predominant.)
- **Safety needs** are when people are striving to find personal security, financial security, health, and well-being. Safety needs are not being met if you are in a war zone, or amidst a natural disaster, or living with abuse or post-traumatic stress disorder (PTSD). After a person has these two basic sets of needs met, then the third level of needs, Love/Belonging, becomes more predominant.
- **Love/Belonging** needs are very powerful and, as evidenced by research in situations of abuse, can even override safety needs (e.g., Stockholm syndrome, children clinging to abusive parents). These are interpersonal and involve our feelings of belonging with friends, family, and lovers. We all know how we feel when we are surrounded by love and feel that deep sense of connection. I feel braver and more capable when I am in a loving relationship with someone, or in regular communication with friends and family who affirm my value.
- **Esteem needs** relate to the need to be respected. Are you valued and accepted by others, and are you valued and respected by yourself? According to Maslow, self-respect is a higher level of esteem than is respect from others. But we can all find ourselves mirrored in others and struggling with low self-esteem. Of course, more extreme manifestations of esteem occur with depression, but social media and the living of life out loud and in front of everyone for judgment can lead to inferiority complexes.
- **Self-actualization needs** refer to what a person's full expressed potential in life can be. Do you have a strong desire to accomplish everything that you can in life, and are you focused on it? Some people have a clear and strong focus on this need—the LeBron Jameses, Richard Bransons, and Oprah Winfreys of this world—who focus energy and attention on being the best and constantly doing the work to improve themselves. Maslow's view was that you couldn't

meet self-actualization needs if the previous needs were not met, but look at Vincent van Gogh—he clearly struggled with love, belonging, and esteem, even though he was very expressed or self-actualized in terms of his art. People may seem successful, even if all their needs are not met. However, dig below the surface and you may discover that there's some deep, unfulfilled need under the veneer of success.

- **Self-transcendence** was a later evolution of the theory by Maslow. He developed the concept of self-actualization to include reaching for the highest levels of human consciousness by extending one's reach to others (more of a Mother Theresa, Nelson Mandela, or Martin Luther King, Jr. model of needs).

THE ROLE OF THE ENVIRONMENT

Make your space a place you like to be in, with fresh flowers, cacti or succulents, essential oil diffusers, candles, comfortable chairs, photos of your favorite people. You DESERVE it!

Are you someone who can't cook until the kitchen is clean? Do you find the noise or smells in your environment affect your ability to concentrate or be productive? Your environment has a huge impact on how you work, how well you work, and what you do. Author James Clear wrote a great blog article about this.[7] In it, he referenced one of my favorite people, Jared Diamond, who I first came across when I studied human evolution and sexuality as an undergraduate and read his book *Why is Sex Fun?* (besides the obvious reasons!).

In his book *Guns, Germs, and Steel*, Diamond points out that the shape of a continent affected how agriculture spread.[8] The east–west shape of Europe, the Middle East, and Asia meant that crops spread easily because the climate was the same, compared to North and South America, where the climate varied massively from the deserts of Nevada, to the Rockies of Colorado, to the wet Amazon rainforest. This meant that agriculture spread two to three times faster in Europe, the Middle East, and Asia compared to North and South America, which in turn meant that the population grew because there was agriculture to support it. Thus, armies developed and technologies and innovations grew. Potentially, this was one of the main reasons why the Europeans rose to power. How freaking cool is this? I love this kind of stuff, it's SO exciting, isn't it?!

---

7 Clear, James. "Motivation is Overvalued. Environment Often Matters More." James Clear. http://jamesclear.com/power-of-environment.

8 Diamond, Jared M. *Guns, Germs, and Steel: The Fates of Human Societies*. New York, NY: W.W. Norton & Company, 2017.

So, what can we do about our environment? Well, we can create an environment that supports our healthy habits. Hide the naughty food in your house (or don't buy it at all), lay out your sneakers and gym clothes by the bed for your morning workout, and use mobile apps to block use of your phone or social media websites during work hours.

Research supports the massive effect our environment can have on our behavior—candy bars are placed at eye level at the checkout and not in the fresh produce section because we are fatigued after making many decisions throughout the grocery store and we are visual beings. Two books that dive into this kind of research more deeply are *Nudge: Improving Decisions about Health, Wealth, and Happiness* and *The Slight Edge: Turning Simple Disciplines into Massive Success and Happiness*. These authors point out how much we are affected (manipulated?) by the cafeteria organization, the Apple store, colors, etc.[9 10]

MOTIVATION: SELF-DETERMINATION THEORY

Can't get motivated? Tidy regularly so your space is organized!

- Set the timer for three minutes or put on your favorite song, really loudly, and get each family member to do one room (e.g., toys away, clothes in laundry/hung up, coat/shoes/bags/sports bags tidied, empty dishwasher!)
- Choose one area, e.g., entry way, linen closet, sitting room, and do ten minutes a day.

The research on motivation is fascinating and multifaceted and there are a plethora of different theories and studies out there.

The self-determination theory of motivation, proposed by Deci and Ryan, is an all-inclusive theory that considers the intrinsic motives (your internal world) and the extrinsic forces (e.g., the environment, other people), coupled with social and cultural elements in determining how we behave in effective and healthy ways.[11]

It's super complicated and academic—broken down into six mini-theories, each of which explains a different facet of human behavior. This little book is not the place for a full discussion of it, but, suffice to say, this theory encompasses all aspects of how

9 Thaler, Richard H., and Cass R. Sunstein. *Nudge: Improving Decisions About Health, Wealth, and Happiness*. London: Penguin Books, 2009.

10 Olson, Jeff and John David. Mann. *The Slight Edge: Turning Simple Decisions into Massive Success and Happiness*. Austin, TX: Greenleaf Book Group Press, 2013.

11 Deci, Edward L. and Richard M. Ryan. "The General Causality Orientations Scale: Self-Determination in Personality." Journal of Research in Personality 19, no. 2 (1985): 109–134. Accessed at: http://www.sciencedirect.com/science/journal/00926566 and http://www.sciencedirect.com/science/journal/00926566/19/2.

we motivate ourselves. The takeaway message is that, based on all the research, we are most effective—read: best performance, most persistent and creative from a free-will motivational standpoint—when we experience the psychological needs of autonomy, competence, and relatedness. That is, *when we are acting from a place of self-motivation and self-drive, we feel capable and competent in our work, and we are connected to others, to the goal, and to the social frame within which we are acting*.

GRIT

> Write a list of three things you have done that demonstrate grit. For example: run a marathon, get a degree, go through childbirth! And then acknowledge yourself for your awesomeness!

This notion, one of many presented for us to consider and lovingly captured by Angela Duckworth in her book *Grit: The Power of Passion and Perseverance*, suggests that passion and perseverance, i.e. our excitement about something and our commitment to pursuing it, are the fundamentals for grittiness.[12] She makes a cogent case for the power of grit in yielding success and that—like your errant six-pack abs or my personal fave, the deep V created by tight lower oblique muscles—you can develop your grittiness.

As with many of these theories, it's always nice to read that we have a modicum of control—we can work on cultivating grit by actively engaging in practices to strengthen that muscle.

WHY ALL THE THEORIES?

I presented these theories, which are by no means an exhaustive list, as a means to get you thinking about how to create the conditions that will support you in this evolutionary path. I want you to unleash your power by developing these great habits. Courage, vulnerability, resiliency, conscientiousness, and excellence are also concepts to consider on this path. Do you overcome fear? Can you handle the shame around trying and not always succeeding? How resilient are you to the inevitable setbacks? How optimistic are you in life? How much do you care about the nitty-gritty of getting things done, and does perfectionism get in the way of your continual drive toward excellence?

---

12 Duckworth, Angela. *Grit: The Power of Passion and Perseverance*. New York, NY: Scribner, 2016.

Much has been written about these notions, and this funky little book is not the place to expound on all of them. These different aspects of motivation will ebb and flow, depending on where you are in the journey or how you feel on any particular day. Part of this process is you figuring it out yourself, by applying these different techniques and then trying them out and striving for your continued success. Are you ready?

De-clutter regularly: it's hard to feel productive when your place has too much stuff everywhere:

- Get rid of one item every day. Put in a box and take to charity/sell when it's full.
- Do the Minimalists challenge: every first day of the month, get rid of one thing, second day of the month, get rid of two things, etc.

HOW TO USE THIS BOOK

I want you to consume this book, devour it, and make it yours! This is a little tool book, giving you some key insights into the main self-care tools and habits that, if you work to put them into place, will make your life more pleasurable and will create freedom from that feeling that you're never enough, that there's never enough time in the day to do all the things you want and need to do!

So, read it through from start to finish.

Then, as you begin to put these practices into place, go back and check out the Summary sections, the What Do You Need to Unleash Your Power sections, the Time Out sections, and the hashtags. Perhaps dive into the Journal section at the back of the book (page 153).

- The Summary sections give you the basic points to consider while you are working on that habit.
- The What Do You Need to Unleash Your Power sections lay out the shopping list or organizational stuff you need to consider.
- The Time Out sections are to remind you that it's okay to freak out, and that what I've been asking you to ponder or do might trigger you, and that's okay.
- The hashtags give you words to consider while you are working on that habit. Post your habit changes to social media using these hashtags, use them with your accountability buddy who's making habit shifts similar to

yours, or stick Post-it notes around your house to remind yourself what you're working on.

- The Journal section allows you to start more deeply reflecting on your inner dialogue around the issue or habit at hand.
- The Recipe section gives you some basic kitchen primers to get yourself into a delicious state!

HASHTAGS #Maslow #SelfDeterminationTheory #Motivation #JaredDiamond #YourBigWhy #BodhisattvaVow #Environment #IntrinsicNeeds #ExtrinsicNeeds #Grit #Courage #Vulnerability #Resiliency #Optimism #Conscientiousness #Excellence #Willpower #PassionAndPerseverance

CHAPTER 2

STRESS

I STRESS, EUSTRESS, WE STRESS: WE ALL NEED SOME STRESS

I have had some stress and suffering, as we all have. In the grand scheme of life, I have found it very easy to get all "first-world problems" with my issues and to feel guilty for focusing on my kinds of problems (hey, I'm not a starving refugee). However, it's important not to discount your own experience and feel guilty for being sad, for being angry, and for processing your experience. The key is not to let your problems define your life!

> Feeling squirrelly? Anxious? Overwhelmed? Lengthen your exhales to stimulate the parasympathetic nervous system which will help calm you down!

My biggest stresses included a few upsets with my three kids: kidney reflux, which required probes and scans and chronic antibiotic use; severe nut allergies requiring epinephrine shots; multiple ear infections and two resultant surgeries; a broken limb and surgery to pin a total break of the humerus. And then we had diagnoses that seemed scary at the time: attention deficit disorder and other concerns requiring occupational therapy, physical therapy, other therapeutic interventions, and school district conversations; and then cancer, which involved multiple surgeries and chemotherapy.

Then, I was deeply affected by my brother-like cousin's death when we were both thirty-seven years old, the death of a friend in college (from an HIV-infected blood transfusion for his hemophilia), and a relationship largely defined by its vituperative quality in relation to me (now ended).

Some events, though "good" or necessary, still generated stress. In a fifteen-year period, I moved seven times, including buying and selling four homes; got married; completed my PhD; emigrated from

London to the USA; and had three kids—all C-sections (don't even get me started on how much this hippie yogini had to work on processing the medicalized births she had) with no family in my city. I learned to navigate long-distance relationships (my amazingly supportive family comes to see me all the time), and I learned how to cultivate and develop my relationships with friends who become family.

> Identify an example of eustress, good stress—the slightly tense gap between where you are and what you want) that you could apply to your personal habits.

This period ended with my ex-husband's difficult but ultimately successful journey to tenure, in which he was rejected by both his department and his second-choice department and eventually got tenure in a third department, which he followed a few weeks later by walking out on us for the fourth time. In retrospect, it was as if he hung on until he got to that point in his career, using my support to get him there! We spent seventeen months going through a painful divorce, and now I am free! We have moved to a place of comfortable co-parenting, living two blocks apart and sharing custody, so our children spend 60 percent of their time with me and 40 percent of their time with their dad.

I love the word eustress—pronounced /yoo'stres/—which is defined as moderate or normal psychological stress, interpreted as being beneficial for the experiencer. For those etymology lovers out there, eu means good and stress means stress, so literally, "good stress." According to Wikipedia:

> Eustress means beneficial stress—either psychological, physical (e.g., exercise) or biochemical/radiological (hormesis). Eustress occurs when the gap between what one has and what one wants is slightly pushed, but not overwhelmed. The goal is not too far out of reach but is still slightly more than one can handle. This fosters challenge and motivation since the goal is in sight. The function of challenge is to motivate a person toward improvement and a goal. Challenge is an opportunity-related emotion that allows people to achieve unmet goals. Eustress is indicated by hope and active engagement. Eustress has a significantly positive correlation with life satisfaction and hope.[13]

13 Wikipedia contributors. "Eustress," *Wikipedia, The Free Encyclopedia*. https://en.wikipedia.org/w/index.php?title=Eustress&oldid=779380024.

So, stress is there. The issue is how to deal with it. There's a great graph, which explains stress and how the right amount can make us productive. We all need a little stress to get stuff done—the Yerkes-Dodson curve (Figure 1) is a great model for this.[14] Along the x-axis (horizontal axis) is stress level, and along the y-axis is performance. You can see that, as the stress level increases, the performance increases, BUT only up to a point. There comes a moment when the stress is so high that the performance level starts to drop.

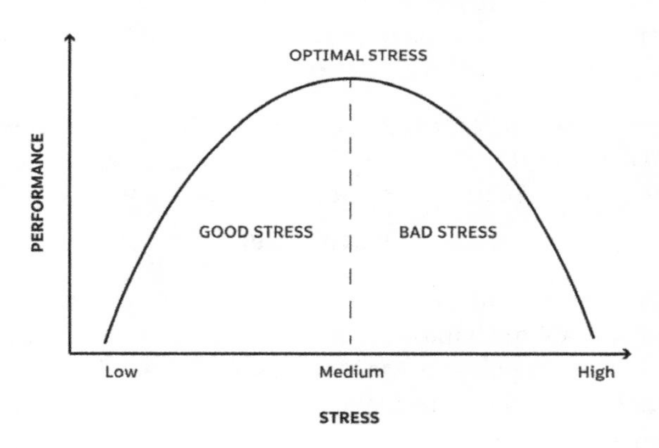

Figure 1. The Yerkes-Dodson Curve

> Feeling unbalanced? Breathe up and down the spine—inhaling up and exhaling down—to help align your chakras (groups of energy—bundles of nerve endings—located along the spinal column).

My point? You need some stress to get a fire under you. But, if you allow stress to take over, then you stop functioning well. Your immune function is compromised, you stop sleeping well, your weight fluctuates, and you might develop diabetes or heart issues. To unleash your power, you need to learn what eustress means for you and how to add a little splash of it to your life, like that perfect amount of hot sauce on your tofu scramble that enhances the experience but doesn't burn your taste buds to ash.

14 Wikipedia contributors. "Yerkes-Dodson law." Wikipedia, The Free Encyclopedia. https://en.wikipedia.org/wiki/Yerkes%E2%80%93Dodson_law.

"The implication is that this basic idea we have that we are controlled by our genes is false. It's an idea that turns us into victims. I'm saying we are the creators of our situation. The genes are merely the blueprints. We are the contractors, and we can adjust those blueprints. And we can even rewrite them."

—Robert Lipton, PhD

OUR STRESS: THE STRESSED NATION

Clearly, I am not alone. According to the American Psychological Association, about three-quarters of Americans are experiencing symptoms of stress (77 percent physical symptoms and 73 percent psychological symptoms).[15]

"The truth is that there is no actual stress or anxiety in the world; it's your thoughts that create these false beliefs. You can't package stress, touch it, or see it. There are only people engaged in stressful thinking."

—Wayne Dyer

I love that quotation by Wayne Dyer. It completely puts the POWER back into my hands. Yes, shitty things will happen in my life. But I have the power to manage this, to unleash MY power by developing great habits which allow me to succeed, manage, and ultimately thrive.

> Feel that fight-flight-freeze system start to fire up? Feeling that cortisol rise? BREATHE. It takes about three breaths to start pacifying the fight/flight and for the cortisol to start dissipating.

You know what stress feels like in your body and mind. Your mouth gets dry, your palms get sweaty, and your heart starts to beat faster. Hormones such as adrenaline and cortisol course through your veins, causing the blood to rush to your limbs for the fight-flight-freeze response. Most problematic in the immediate term is the fact that blood loss occurs in your brain and digestive tract as the blood gathers in your limbs. So, your ability to make good decisions is compromised and your ability to process food well is impacted.

Adrenaline breaks down over a matter of minutes. Cortisol, however, takes twenty-four hours to break down, which means that, if you have one stressful moment every day, you are essentially living in

15 "Stress a Major Health Problem in the U.S., Warns APA." American Psychological Association. October 24, 2007. http://www.apa.org/news/press/releases/2007/10/stress.aspx.

a cortisol bath. This indicates that if you are chronically stressed, your body is storing fat in your belly, your immune function is compromised, and, if you're a woman, your body starts using progesterone to make cortisol (yes, progesterone is a precursor to cortisol), as noted by Dr. Claudia Welch in her book Balance Your Hormones, Balance Your Life. This means your ability to have healthy cycles, conceive, maintain a pregnancy, and so on is hugely compromised. She writes:

> In the modern world, many of us find ourselves in a near-constant state of stress. When this happens, we have stress hormones like cortisol coursing through our bodies on a regular basis. The crunch comes when we don't have enough of these hormones available to satisfy the huge demand for them. Then our bodies will find a way—any way—to get us more. One way to get more is to sacrifice some of our sex hormones by transforming them into stress hormones... Progesterone can become cortisol. Good-bye Fred Astaire; hello Fred Flintstone.[16]

Testosterone—which yes, women have as well as men (and men have progesterone and estrogen)—is also implicated here. Low levels of testosterone, which are found in highly stressed people, cause weight gain around the midsection, and depression. Cortisol and testosterone operate together. Interestingly, super successful executives have high testosterone and low levels of cortisol.

> If stress is stuck in your body, moving can be really helpful. Don't overthink it—boost those endorphins so you can re-integrate your mind and body!

I am a fan of the super quick way to unleash my masculine energy and boost testosterone, which involves doing one of Amy Cuddy's three so-called power poses: Wonder Woman (hands on hips, feet 18 inches apart), I Won (arms raised over head, hands in fists, and feet apart), or Chief Executive desk pose (sit back in your chair, hands behind your head and your feet on the desk).[17] Interestingly, this data has not been replicated, and some are now disputing it, but I still use it, because I find myself feeling more powerful and "in charge" when I do these poses! Ahhhh...the power of the body-mind connection!

---

16 Welch, Claudia. Balance Your Hormones, Balance Your Life: Achieving Optimal Health and Wellness through Ayurveda, Chinese Medicine, and Western Science. Cambridge, MA: Da Capo Press, 2011.

17 Cuddy, Amy. "Your Body Language Shapes Who You Are." TED Talk. June 2012. https://www.ted.com/talks/amy_cuddy_your_body_language_shapes_who_you_are.

Anyway, back to stress. I am not discounting the mighty power of stress hormones. We all know how productive they can make us. The issue is whether you are dealing with chronic, high-level stress and the resultant disruption of your systems.

EVOLUTION AND STRESS: NATURE AT WORK

Now, it's not surprising that we are stressed out when you look at evolution, rather than just looking at stress on a specific personal level. The world is evolving at a speed that is beyond our human ability to grasp and react to, at least in the way that we need.

Humans have been around for tens of thousands of years. During the majority of that time, we rose and slept according to the cycles of dawn and dusk, we ate what was available to us—either by hunting or gathering—and exercise was a natural part of our day.

We ate locally, which included much less sugar, salt, and fat compared with today's diets, and no processed products. Our relationships were shorter, due to shorter life expectancy, and if you have studied or read any evolutionary psychology, you'll know that there were a huge variety of different models for relationships and child-rearing, not just the monogamous model that is promoted as being the norm and is dominant in the west today.

Think about the last one hundred years. We now move around the world and beyond (planes, rockets, trains, boats, cars, Rollerblades) at a speed that is unprecedented. Information moves around the world at a previously unknown speed and is being presented to us at such a high rate and density that we can't possibly take it all in. Consider how humans have changed the world with the Industrial Revolution in the 1800s and then with the advent of major technological changes during the last 30 years—the Internet, smartphones, medicine, space travel, etc.

Is it really surprising that our bodies and minds have NOT caught up?

EPIGENETICS: NURTURE AT WORK

How we eat, sleep, work, connect with others, and consume, share, and explore information has changed at a speed that is too fast for the human mind-body to keep up with. What does this mean? We have no clue whether what we are doing is having far-reaching effects on our mind-body. Epigenetics, a new field of research, shows that, although we may have a genetic proclivity for certain diseases or disorders, how we live—the food we eat, the exercise we do or do not take, the alcohol/drugs/cigarettes we consume, and whether or not we meditate—can markedly affect how our genes are expressed.

"Genes load the gun, but environment pulls the trigger."

—Robert Lipton, PhD

We are not slaves to our genes. We can potentially prevent 80 percent of chronic conditions by the DECISIONS we make on lifestyle! This is massively emboldening, don't you think? We can unleash our power by developing the amazing kinds of habits that DON'T pull the trigger!

HOW DO YOU HANDLE STRESS?

> Essential oils are mighty! Put in a diffuser, or add to a small bottle with distilled water or rose water and a pinch of Epsom salt, for example, vanilla, frankincense, lavender, or vetiver for grounding. Citrus oils for creativity. Peppermint and orange for energy!

I shared a home for many years with someone whose response to stress was to eat more, drink more, smoke, stay up late, sleep late, work harder but not smarter, and withdraw. I tried that approach and I noticed that it did not make me feel better. In fact, I felt worse—much worse. I felt sludgy and slow, and my skin was grey. I used caffeine and sugar for energy because I did not have innate energy.

As my insomnia blossomed and I started struggling with food (bloating and irritable bowel syndrome-like symptoms), western doctors said: take sleeping pills, get tested for sleep apnea, let's test you for celiac disease, let's test you for Helicobacter pylori (to see if the bacteria, which about half of us have, were causing ulcers)...let's test, test, test.

"Ugh," I thought.

I was in my thirties and my body and mind used to be able to handle this. I wondered, why are you doctors suppressing my symptoms and not giving me a helpful framework to really help me figure out how to manage this?

I read *Perfect Health: The Complete Mind/Body Guide* by Deepak Chopra, MD, and learned about the Ayurvedic system of doshas: the three parts which make up all of our constitutions.[18] We all have three doshas, but we vary in how much of each we have, and environmental (psychological, emotional, seasonal, situational) events will impact whether we have too much of a particular dosha at any one time.

It was the first time I had come across a classification system that blended mind and body (unlike the nine-sided Enneagram test, Myers-Briggs Type Indicator, or Body Mass Index). It also acknowledged that what works for me might not work for you. I started doing the dosha tests on my family members and noticing themes and patterns. This approach to engaging with the world and my own unique mind-body system was starting to really make sense to me. I started to feel the rumble of personal power being unleashed. I could take back the reins by developing great habits!

In 2011, I went to a workshop run by the Ayurvedic practitioner James Bailey and did a private consultation with him, where he told me to be careful about being vegan—that I needed to ground myself with some heavier oils inside and outside my body. I started my *abhyanga* practice (warm oil self-massage) and really learning to listen to my body and what it needed. With clear prescience, he also saw my ex-husband and told him that he was concerned about our marriage.

Eighteen months later, I went to an Ayurveda workshop for yoga teachers, run by Cate Stillman, and committed to a nine-month certification in Ayurveda. As I mentioned in Chapter 1, it is the sister science to yoga, and is built around this idea of daily rhythms and how you must start by looking at the context of your life, rather than starting with symptom suppression.

My life changed forever: I realized that to create freedom, I needed to create routine. Routines for myself—not just for my kids and my yoga students—that, rather than being restrictive, would generate an unbelievable level of freedom. I finally understood—that to unleash my power, I needed to develop great habits. And that these habits, when sequenced well, when nurtured for my unique body-mind combination, would allow me to become the best version of myself, a.k.a. a freaking goddess rock star!!

---

18 Chopra, Deepak. Perfect Health: The Complete Mind/Body Guide, Revised and Updated Edition. New York, NY: Three Rivers Press, 2011.

A BETTER MODEL FOR UNDERSTANDING AND HANDLING STRESS

One of the things that I find difficult (er, stressful!) about some of the experts out there who tell us what to eat, how to exercise, and so on, is that there is often a one-size-fits-all model.

Ayurveda presents two main approaches to managing the daily rituals and self-care in your life: the general and the specific. The general connects us all—what the season is, what time of day it is, what age we are—in terms of the best approaches to being happy and healthy and deeply well.

The specific acknowledges that we all have different constitutions: both physical and psychological/emotional, which respond in a variety of different ways to what's happening in our physical and psychological/emotional world. These variations mean that we respond differently to what's happening to us and, thus, how we manage ourselves will vary from person to person, depending on our own unique constitution.

DOSHAS: ORGANIZING ENERGETIC PRINCIPLES

Our world is made up of five elements: space, air, water, earth, and fire. These are grouped into three organizing energetic principles, the doshas, which form our lives, our seasons, our days, and our mind-body.

- **Kapha** (water and earth) controls structure, joints, lubrication, and lungs. Kapha is associated with grounding and soothing energy, bones, stability, growth, and reliability. Kapha types tend to be warm and steady personalities who exude an "earth mother" energy. When they are imbalanced, Kapha people tend toward lethargy, oversleeping, withdrawal, depression, and weight gain. Kapha people have slow metabolisms (can manage eating two meals per day) and are attracted to dense foods (similar to earth element) and thrive better in dry weather (opposite to water element).
- **Pitta** (fire and water) controls digestion, metabolism, and energy production. It is associated with heat, oil, intensity, leadership, and strong digestion. Pitta people tend to be outspoken, fiery, and great leaders. When they are imbalanced, they get rashes, ulcers, and skin disturbances. They have very short tempers and tend to have explosive anger. Pitta people

will thrive better in cooler weather (opposite to fire element) and be attracted to spicy, caffeinated, or alcoholic foods and drinks (similar to fire element).

- **Vata** (air and space) controls movement in the body and the mind—blood flow, breathing, nerve energy, and the elimination of waste—and is responsible for about 70 percent of the diseases we get. Vata is associated with cold, dryness, movement, creativity, mental quickness, lightness, and nerve energy. Vata people tend to be energetic and creative with a strong physical presence. When they are imbalanced, they get nervy, have disturbed sleep and digestion (constipation), and are prone to anxiety. Vata types are attracted to light, dry (astringent), and airy foods such as popcorn, raw cold salads, and apples (similar to air and space), and will thrive better in warm, humid environments (opposite to air and space element—inherently cold).

When we are born, we are in the Kapha stage of life—we grow quickly, sleep a lot, and are prone to sinus and/or ear infections (watery element of Kapha). Then we move into the Pitta-dominated teen and adult years and things start to get fiery—acne, hormones, testing boundaries, anger, exploring leadership, and taking our place in the world. Then, in middle age and old age, we move into Vata, where we need less sleep, our skin gets dry, we need less food, and we become less connected to the physical world.

The fall is associated with Vata—movement, wind, change. Winter and early spring are associated with Kapha—plants return to the ground and gather energy for growth (it's cold and wet). And late spring and summer are associated with Pitta—heat.

Our days cycle around this clock, too (Figure 2): 10:00 a.m.–2:00 p.m. (Pitta time), 2:00 p.m.–6:00 p.m. (Vata time), and 6:00 p.m.–10:00 a.m. (Kapha time).

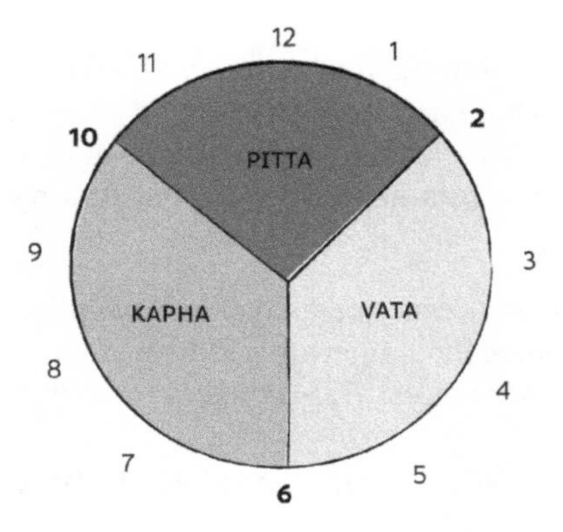

Figure 2. The Ayurvedic Clock

Again, we all have these three doshas in our system. The variation is what makes us unique. Most of us have one dosha that is dominant and then a second that is not far behind. Very few of us are truly tri-doshic. What this balance—your constitution—will tell you is what you crave and what you need to maintain a calm and happy life.

So, for example, if your primary dosha is Vata, you will resist routine, desire light foods, have a tendency toward insomnia and anxiety, and have a tendency towards dry skin and constipation. Even though everyone tells you salads are good, and you will crave these light foods if you are high Vata, eating lots of raw cold veggies in January, when it's cold and your system is struggling to stay warm, is not doing yourself any favors. What you need is the opposite to soothe your system. A bowl of warm, moist cooked veggies, drizzled with oil and served with avocado, is going to be much more nourishing. Generally, no one should eat cold salads in the winter, but if you're high in Vata, it will be worse. You got it?!

Now, don't go getting all obsessed on me. Start with general rules—that's the best way to develop great habits. What season are you in? Winter? Eat warm roasted vegetables, soups, etc. Summer? Eat salads and colder, lighter foods. What time of the day is it? Lunchtime? Eat your biggest meal because it's Pitta time, which is associated with higher bile, and gravity helps with digestion. Evening (Kapha time)? Eat light and easily digestible foods.

SUMMARY

Stress will do bad things to your mind and body—the world has evolved at such an extraordinary rate, it's not surprising we haven't caught up! So, start to pay attention—how you live will change how you do. Your genes are not your destiny. Learn how to manage yourself!

Consider the doshas—what time of day is it, what season are you in? Use these three energetic frameworks to help guide your sleeping, eating, and exercise choices. This is the key to unleashing your power!

WHAT DO YOU NEED TO UNLEASH YOUR POWER?

If you're so inspired, do three dosha tests (free online) to get an overall picture of your constitution. You are born with a particular balance (your prakruti) and what you are right now is your vikruti, which may be out of balance due to your physical and psychological/emotional worlds right now. If not, just start to notice the seasons and what you're doing to support yourself in this season. Start to construct frameworks to support your constitution—which I discuss in depth in the next chapter.

TIME OUT

I know. Lots of new information—but that's empowering, don't you think? Stress can be managed, but you have to be willing to own it, by reflecting on your own life and then how you manage your days. Are you taking time off? Do you schedule space between your appointments? Do you take vacations (e.g., a screen Sabbath on the weekend, or a weekend off per month, or a week off two to four times a year)? How do you react when people lash out? What are you eating, reading, and watching? Who do you spend time with? How do they make you feel?

HASHTAGS #Eustress #GoodStress #YerkesDodson #Stress #FirstWorldProblems #Adrenaline #Testosterone #Cortisol #Progesterone #FightorFlight #PhysicalStress #PsychologicalStress #Evolution #TechnologicalRevolution #MeditationImprovesImmunity #Epigenetics #Lifestyle #NurtureVersusNature #StressResponse #Insomnia #DigestiveIssues #Ayurveda #Routine #FreedomThroughRoutine #Doshas #RefuseVictimhood #Motherhood #Yoga #Pitta #Kapha #Vata #EatWithYourSeasons #EatBigLunch #Constitution #Balance

CHAPTER 3

UNLEASH YOUR POWER: DEVELOP HEALTHY SLEEPING, EATING, AND EXERCISE HABITS

The secret to unleashing your power is developing great habits. There are lots of great habits to develop. So, after working with myself (for over seven years) and my clients for over five years, I zoned in on a few key habits.

HEALTHY

Healthy means something different for everybody. For you, healthy might look like eight hours of sleep, mostly raw food, thirty miles of running every week, cool and dry climates, and a busy social life. Or it might involve a well-cooked and stew-based diet, seven hours of sleep, gentle yoga every day, weights a few times a week, warm and humid climates, and quiet nights at home.

We all have different notions of what healthy means. The Merriam-Webster Dictionary defines healthy as: "in good health; free from disease or pain; beneficial to one's physical, mental, or emotional state." I prefer the Ayurveda definition for healthy, which is *swastha* and means "seated in the self," that is, all parts of yourself—your mind, your body, your soul–are connected and functioning well.

> Healthy is a loaded term. Start paying really close attention to how you feel when you do or don't do certain things! Write it down. Think about it. Become intentional.

We are all biased as a result of our education, what our parents taught us, or what we watch on TV, for example. Many of us were taught that we needed to drink cow's milk to have healthy bones and that we needed to consume large quantities of animal protein to have healthy, strong muscles. Others were taught that fat is bad. Some of us grew up with the idea that regular snacks and lots of little meals were the best, or that we should catch up on our sleep at the weekends. And then there's all the exercise concepts around "healthy." And so on and so forth. This puts us in a somewhat difficult situation. To compound the issue, we are all unique, and I don't mean it in the oft-ridiculed snowflake-generation way. I mean it in the "our mind-body systems are all built differently" way, as I explained in Chapter 2 in the section on doshas.

Ever noticed that some people love heat? Others hate it. Some people love crowds; others like solitude. One friend loved being pregnant; another hated it and felt sick and heavy. Another friend could run for thirty miles, while yet another could barely make it through a quarter mile. Some people can eat anything and poop like a champion; others get the runs or get constipated at the mere sight of a bowl of chili, popcorn, or the variation in routine brought about by an airplane and time change. This difference in constitution is what makes you unique.

This uniqueness is fantastic. But don't get too hung up on this—I want you to start by following the general guidelines discussed next. As the old saying goes, you've got to learn how to follow the rules before you can break them. I know, seems super boring, right? But this is what will set you free, and I know that's really why you're here.

HABIT

We live by habits—over 45 percent of our daily actions are habits. The issue is whether these are good habits—ones that are serving you and the kind of life you want—or not. I help people create connections between the habits that are working, add those habits that make their lives better, and replace habits that no longer work.

I like to think of myself as a habit scientist—your Chief Habit Scientist, a.k.a. someone who helps you cut through the crap that is no longer serving you. And I want you to be your own habit scientist, too—someone who really starts to pay attention to yourself. How do you feel when you go to bed at 10:00 p.m. versus midnight? How do you feel when you eat chocolate and drink caffeinated soda and text in bed until 2:00 a.m. versus going to bed on an empty, de-caffeinated belly without your phone? Did you sleep better when you exercised?

"Excellence is an art won by training and habituation. We do not act rightly because we have virtue or excellence, but we rather have those because we have acted rightly. We are what we repeatedly do. Excellence, then, is not an act but a habit."

—Aristotle

So why do this, you might ask? You make tens of thousands of decisions every day. By creating connections between habits, you remove the decision. If you remove the decision, then you free up your mental space, time, and energy for the good stuff! This is important to do because willpower won't work forever, which is why diets never last in the long term.

Willpower is a muscle—and, like our own natural constitutions, we have varying amounts of it. Willpower can be trained, but it does get fatigued. My approach to creating connections between habits essentially works by reducing the willpower element because you start to connect your new habits with current habits that are serving you.

Do you remember learning to drive a car, or learning a new route to get to work? You start by thinking through all the steps: check that gears are in neutral, place your foot on the brake and clutch, turn on the ignition, check the mirrors, turn on your indicator, release the parking brake, put the car in gear. Accelerate, depress clutch, move into second gear…etc. After a while, you stop thinking about all those micro-decisions and just do it without thinking. That's the goal here—make routines that you don't have to think about, and then life is free!

"Early to bed and early to rise makes a man healthy, wealthy and wise."

—Benjamin Franklin

Essentially, my approach to these healthy habits is all about integration and connection and how that creates freedom. It's all about connection—connection between your thoughts, words, and actions; connection between what you think, say, do and feel about all of this; connection between the different layers of yourself (what's called the sheaths or koshas in yoga and Ayurveda—your physical body, your mental body, your breath and energy body, your intuitive body); and connection between yourself and others (which is enhanced, developed, and deepened if you start by working on yourself!).

"I think there's some connection between absolute discipline and absolute freedom."

—Alan Rickman

This quotation by the late, great Alan Rickman, to my mind, totally nails this issue. To fully unleash your power, which means for many of us to live with a deep-seated sense of freedom (from disease, pain, distress, anger), developing a rigorous set of great habits is key!

WORKING ON YOU INCLUDES CARING FOR SELF: FILL YOUR CUP

Healthy is about you finding what works for you and learning to work with the days, seasons, environment, and your life. As I mentioned in Chapter 1, it's super easy to allow others to drink from your source—draining you of time, energy, and your va-va-voom for life—so this work is all about learning to put YOU first. It's about filling your own cup first, and letting it overflow with abundance onto others.

I was initially very uncomfortable with the concept of self-care, because it seemed selfish to me. That was because one of the most important relationships in my life had been with someone who did everything from their point of view, on their terms, and I rabidly did not want to be selfish or narcissistic!

What is one small act—something indulgent—that you can do for yourself?

However, my understanding of "selfish" fundamentally shifted when I started to realize that I could not show up and be the best I could in all my other roles (wife, mother, teacher, friend, coach, etc.) if I did not take care of myself first. Selfishness is like noonish, according to a friend and healer who said to me, "Selfishness is kinda, sorta about you—like how 'I'll call you at noonish,' means 'I'll call you around the noon hour.' It's not narcissism." It's not: everything is from my perspective and mine only (which implies a lack of empathy). It's about me, and me taking care of myself, so that I can be in connection with you in the best and healthiest way possible. It's not a rigid, me-first approach. Rather, it's the idea that I'll fill my cup so that you are not draining me, but are sipping of the abundant overflow. You get my drift?

Women, mothers, and caregiver types struggle with this concept quite a bit—putting our needs first, so that we can show up as ourselves.

And yet this concept, this very necessary self-care, is sooooooo important. If we are to truly unleash our power in the world as mothers, lovers, creators, writers, thinkers we need to develop the great habits that will support us.

I found that a helpful way of thinking about this was to start to learn about myself not in relation to others. I am a great coach, daughter, friend, mother, and lover. I listen, I care, I connect, I hold space for you, and I help with your needs.

But for many years I lost myself—putting the roles of wife and mother ahead of me, the person. So, I had to learn how to put my needs first—I had to learn to put on my own oxygen mask first (as I noted in Chapter 1). This was really tough, I'm the first to admit!

The other reason I was averse to this way of thinking was that routines and rituals sounded middle-aged. Planning my groceries, bulk cooking, and having a bedtime seemed containing and boring and I am anything but. I like to be free and wild (I'm high Vata, which is the air/space dosha, so freedom is part of my constitution!) and fly by the seat of my pants. What I was failing to see at this point was that, by containing myself, by creating rituals and routines, I was creating freedom. By connecting habits around my daily activities, I was removing the decision-making process, and by removing the decision-making process, I was freeing up my mind—my mental hard drive, if you will—so I had more space to engage in what I wanted!

Don't divorce yourself from your life!

THE SCIENCE OF HABIT—IN BRIEF!

Habits are made up of three main parts: the cue, the action (habit), and the reward. The cue and the reward are easier to access because they are occurring in the frontal cortex (the most evolved part of our brain).

Did you know our brain evolved up and forward? The brain stem evolved first and is responsible for the actions that keep us alive, such as breathing, and all those internal functions that we don't think about, such as peristalsis (moving goop along the intestines). The limbic system evolved second and includes parts like the hypothalamus, hippocampus, and amygdala, which are involved with making memories and processing fear.

The most advanced part of the brain is the frontal cortex, which evolved in tandem with these other parts into a complicated and extraordinary group of neural networks, which govern our behavior and thoughts and feelings and so on. Okay, I'll stop nerding out, I can tell you're drifting off...the point is, it's easier to think about and change the cue and the reward than the action, because we have more access to our frontal cortex.

The Three S's of Cultivating a New Habit:

Same time, Same place, Sandwich. If you want to add a new habit, do it at the same time, in the same place, and sandwich it between two habits that are already strongly linked. Example: Want to start meditating? Sandwich it between two things you always do consecutively, like let out dog/make coffee, or brush teeth/shower.

So, take the Charles Duhigg example from his fantastic book, *The Power of Habit.*[19] You put on ten pounds. You reflect and realize that at 3:00 p.m. every afternoon, you notice you're going to get a cookie. Time to take charge of your own habits by accessing the cue and the reward part of the habit cycle and by being your own habit scientist.

What is the cue and what is the reward? Break it down. Change the variables. So, maybe you're getting up from your desk at 3:00 p.m. every day because you're hungry and your blood sugar has dropped—so try eating an apple instead and see how that feels. Or maybe you're in need of some exercise—you've been sitting on your butt for three hours and you need to get the blood flowing! Try going for a walk. Or perhaps you want social connection—try getting up and going to find someone to converse with! See? By varying and working with different cues and noticing how the rewards feel, you're accessing the bits you can, in your frontal cortex, and experimenting with what you're really hungry for!

We all know this from seeing people we know and love or admire from a distance struggle with addiction. Right? The people who have the most success in creating big changes in their lives are those who:

1. **Replace their unhealthy habit** with a healthy habit, e.g., replace drinking with Alcoholics Anonymous meetings.

2. **Have social support and accountability**. This is SO vital. Finding a group of people to be a part of who are also experiencing what you are going through is key. Research

19 Duhigg, Charles. *The Power of Habit: Why We Do What We Do in Life and Business.* New York, NY: Random House, 2014.

shows that people who live the longest—in the so-called blue zones of the world, as author Dan Buettner describes in *The Blue Zones: 9 Power Lessons for Living Longer From the People Who've Lived the Longest*—have social connectedness, strong family orientation, healthy eating habits, daily activities, and no smoking.[20]

Research looking at aging, disease, and physical health shows, for example, that social disconnectedness and the perception of isolation are both associated with lower levels of self-rated physical health. Further, people who are not receiving strong social support are 50 percent more likely to die from illness than those with support, and those with chronic conditions thrive (make doctors' appointments, remember to take blood pressure, eat better, etc.) when they have social support.[21 22 23]

More recent research, published in the gold standard of journals in this field—science—shows that poor relationships or a lack of social connectedness, which is part of the Love/ Belonging Needs discussed in Chapter 2, significantly impacts your long-term health.[24]

3. **Let go of who they think they should be** and be the person they were destined to be (ideal self). Or, stop "must-er-bating" and just be!
4. **Let go to something bigger than themselves**, such as God, the Muse, the Universe, collective energy.
5. **Remember that these shifts take time**, so they are gentle and kind to themselves (and nonjudgmental of those who are struggling, especially when there are relapses).

20 Buettner, Dan. *The Blue Zones: 9 Lessons for Living Longer From the People Who've Lived the Longest*. Washington, DC: National Geographic, 2012.

21 Cornwell, Erin York, and Linda J. Waite. "Social Disconnectedness, Perceived Isolation, and Health among Older Adults." *Journal of Health and Social Behavior* 50, no. 1 (2009): 31–48. https://www.ncbi.nlm.nih.gov/pmc/articles/PMC2756979/.

22 Blue, Laura. "Recipe for Longevity: No Smoking, Lots of Friends." Time. July 28, 2010. http://content.time.com/time/health/article/0,8599,2006938,00.html.

23 Gallant, MP. "The Influence of Social Support on Chronic Illness Self-Management: A Review and Directions for Research." *Health Education and Behavior* 30 (2003): 170–195.

24 House, J., K. Landis, and D. Umberson. "Social Relationships and Health." Science 241, no. 4865 (1988): 540-45. http://science.sciencemag.org/content/241/4865/540/tab-article-info.

"Sow an act and you reap a habit. Sow a habit and you reap a character. Sow a character and you reap a destiny."

—Charles Reade, nineteenth-century novelist and dramatist

I know, it can seem funny to do this, but learning to take charge of yourself and your world is what will make you happy! When my life was turned upside down, this stuff worked so well for me. It brought me a deep sense of connection with myself. So I am beyond excited to share it with you. Being the nerd girl I am, I trained, thought, practiced, and then started teaching this to people and coaching people to help them get organized.

HABIT CULTIVATION IS TOUGH AS SHIT!

Accountability If You Live Alone:

Take a moment to reflect on a key habit you want to change. (a) Who do you know who also wants to change that habit? Can you be accountability partners? Construct a plan—remember, small and easy, e.g., meet at yoga on Monday evenings and at the pool on Thursday evenings. (b) Hang a calendar, check off the dates when you succeed.

I do want to note here that I know this stuff is really hard, which is why people like me exist—coaches to help you learn how to keep accountable to your path. One of my favorite teachers, Pema Chödrön, mentions, in her book *Comfortable with Uncertainty: 108 Teachings on Cultivating Fearlessness and Compassion*, that we often use three main strategies to rationalize our crappy behavior when we notice that we have fallen into bad habits, such as laziness or anger (which, believe me, will come up for you when you start doing this massive, life-changing habit evolution work!).[25]

The first is attacking: we shame ourselves for hitting the snooze button (rather than getting up early and meditating and exercising) or for eating that doughnut or cookie. Then, we wallow in guilt and the feelings of badness!

The second is indulging: we condone our behavior: "That's just the way I am," "I deserve this cookie," "I need to sleep for fourteen hours over the weekend." Inadequacy and self-doubt may plague us, but we become really good at rationalizing or condoning our shoddy behavior.

The third is ignoring: we space out, dissociate, or compartmentalize. This can be effective for a while, but usually something happens that

25 Chödrön, Pema. *Comfortable with Uncertainty*: 108 Teachings on Cultivating Fearlessness and Compassion. Boston, MA: Shambhala, 2003.

will point out this behavior. Either we get one of those massive in-your-face moments (a diagnosis, a relationship ending, a job loss), or we get a nagging sense that things are not good and we could do a whole lot better with our lives.

This is really important stuff to figure out.

KAIZEN

> Kaizen—what is one small change that you can make, that will produce a massive effect?

This is a fantastic Japanese concept that in recent years has been applied in coaching settings. The key point is to make the smallest possible change, continuously, and commit to it "as if the devil and his pitchfork are chasing you," as author Steven Pressfield puts it.[26]

For example, do you have a deep sense that meditation is needed in your life? Then make a small, doable shift to make it happen. Commit to doing it every day, after you rise and before you brush your teeth, for five minutes. Just sit on the floor in the hallway between your bedroom and your bathroom and do it. Then, after you're repeatedly doing it for a few weeks, up it to ten minutes and so on. Make the change so small and easy to achieve that you can't possibly fail! Consistency to the habit is the key.

RELATIONSHIPS

> **Accountability If You Live With Someone or a Family:**
>
> Take a moment to reflect on a key habit you want to change. Does the person or do people you live with also want to change that habit? Can you do it together? Create a simple, easy plan. Example: Twice a week bulk cook for packed lunches, so you're not eating unhealthy cafeteria gruel/fast food.

This is about two main relationships: your relationship with yourself and your relationships with others. Getting clarity on how you view yourself and your strengths and weaknesses, and whether you truly face yourself and your life, are the keys to being a happy and successful individual. Becoming clear on your boundaries and expectations in your relationships with others, and how these relationships ebb and flow as a result of your habit evolution, must be faced if you're going to navigate successfully.

---

26 Pressfield, Steven. *Do the Work!: Overcome Resistance and Get Out of Your Own Way.* The Domino Project, 2011.

THINGS TO KNOW BEFORE YOU START

When you start to work on yourself, your daily habits—who you are—will change. Your identity will shift because your habits are just a reflection of your current identity. So you have to believe the new version of yourself is possible before you even get there yourself. Which is why we have coaches. That's when I became successful—when I got a coach.

She could see where I could go and who I could become before I could see that in myself. Through the coaching experience, I gain clarity on what tools work for me (by being given different ones to try out and report back on), and this helps me see the areas that I might be blind to, so I can shine a light there and start to try and make some shifts in how I work and live.

This is the "hero's journey," and you are the hero (or heroine).[27] Like all the great myths and gods of the ancient world, the hero's journey will involve you facing yourself intimately, struggling and crashing, getting back up again, and answering the call to adventure. Often you need a mentor (in modern-day life, that is the coach, therapist, or evolutionary group) who inspires you to succeed. Just don't give up. You've. Got. This.

SUMMARY

Healthy is different for everyone. Creating connections between your habits removes the decision-making process, which frees up your mental hard drive to think about and do the stuff that really floats your intellectual boat. The main habits I focus on are—sleeping, eating, and exercising, which I discuss in the following chapters. Getting clarity on your relationship to self and other will help you navigate this journey successfully.

WHAT DO YOU NEED TO UNLEASH YOUR POWER?

A large glass of gin. Just kidding! Some time to reflect. Some time to consider what changes you need to make your life successful, some time to talk to those people who are most intimately involved in your life, and then, the tools which I am going to provide you with in the following chapters!

---

27 Campbell, Joseph. *The Hero with a Thousand Faces (The Collected Works of Joseph Campbell)*. Novato, CA: New World Library, 2008.

TIME OUT

The piece that freaks most people out is the notion that by changing habits, you change who you are. Now, don't freak out on me here—you can do this! I just wanted you to know that, and that with awareness comes power, and with that power comes the strength to really do this!

HASHTAGS #Change #YouGotThis #Kaizen #ReduceItToTheRidiculous #FillYourCup #SelfCare #HabitScientist #Habits #BrainScience #Neuroscience #CueActionReward #ChangeTheCue #ChangeTheReward #Healthy #Relationships #SpecialSnowflakes #UniqueConstitution #Unique #PayAttentionToYourself #Willpower

CHAPTER 4

SLEEP

"Sleep is that golden chain that ties health and our bodies together."
—Thomas Dekker

WE ARE SERIOUSLY SLEEP-DISTURBED, MY DARLINGS

Did you know that the United States is in quite a state when it comes to sleeping? About 34 percent to 44 percent of adult Americans fall asleep unintentionally every day.[28] Those are people who might be driving a school bus, operating heavy machinery, or cutting you open in surgery. Sleep deprivation also manifests in different ways. Have you ever found it hard to concentrate, remember things, do your work or hobbies, or sort out your financial affairs? These are all sleep-related difficulties, according to the CDC in 2015.

Keep a sleep diary. Note down:

- When you slept;
- How you felt on rising; and
- What three things you did before you went to bed (e.g., watched TV, had sex, shouted, stressed about finances while paying your bills, drank wine).

In fact, the CDC also reported that 50 to 70 million people have some form of disturbed sleep. The use of chronic prescription sleeping pills has been connected with long-term health issues, as well as negatively affecting those around us—for example, sleeping pill-induced car crashes and the subsequent lawsuits and hospitalizations.[29] [30] Jeez, what a mess we are in!

It's not really surprising that our sleep is insanely disturbed. If you think about human evolution, we (Homo sapiens) have been around for about two hundred thousand years, with recorded civilizations being around for about six thousand years.

28 "Insufficient Sleep Is a Public Health Problem." Centers for Disease Control and Prevention. September 3, 2015. http://www.cdc.gov/features/dssleep.
29 Fox, Maggie. "Sleeping Pill Use Raises Car Crash Risk, Study Finds." NBCNews.com. June 11, 2015. http://www.nbcnews.com/health/health-news/sleeping-pills-raise-car-crash-risk-study-finds-n373891.
30 "FDA Says Pills Can Cause 'Sleep-Driving' " *The Washington Post*. March 15, 2007. http://www.washingtonpost.com/wp-dyn/content/article/2007/03/14/AR2007031401027.html.

During this time, we rose and slept with the natural light. Then, industrialization started in the 1800s—mechanizing, factories, gas lamps, and electricity. We began being able to stay awake and do things outside of nature's hours! Ever tried reading by candlelight or a paraffin lamp? It's nowhere near as bright as a one-hundred-watt bulb. Consider how the world has changed in the last century, particularly in the last twenty-five years. Notice how bright your smartphone is?

In the western, developed world (though I could argue with you about whether the word "developed" is right, but I digress), my generation (Generation X) is the last generation to grow up without the Internet and smartphones and know a world without it. Let that sink in for a moment.

Research shows that gratitude before bed improves sleep—try it! Say it to your bedmate, your kids at bedtime, or write it down!

Our mind-body has not yet evolved to deal with the amount of information, the number of decisions, and the pure level of connection with so many people, things, and ideas that we throw at ourselves every day. I'm surprised we are not a total mess. What's really interesting to me is that this is not a new phenomenon, as Google revealed to me one snowy, cold morning in Cleveland in January of 2017!

Seneca, the Roman Stoic philosopher, statesman, and writer was concerned about the number of books being published in the first century. During the Renaissance (1300–1700 AD), when the printing press was invented (around 1450), scholars started getting concerned about the amount of information that was beginning to be disseminated. How could our poor brains possibly take in all this information? As discussed in Orin Edgar Klapp's book *Inflation of Symbols: Loss of Values in American Culture,* the German sociologist Georg Simmel, who was fascinated with how everything interacts with everything else, talked about this around 1900.[31] Simmel became concerned that city dwellers who had to shield themselves from "indiscriminate suggestibility" would be totally overwhelmed by the amount of information that came their way, resulting in a blasé attitude and jaded people. Can you imagine what he would have made of Times Square in Manhattan or the Shibuya Station Hachiko exit in Tokyo (both of which I have been jostled at, and, whoa, the

31 Klapp, Orrin Edgar. *Inflation of Symbols: Loss of Values in American Culture*. New Brunswick, NJ: Transaction Publishers, 1991.

lights, the mass of humanity!)? These issues have continued to blossom, and we now deal with ridiculous amounts of information.

We have to learn to filter the information and organize it so that we can relax and sleep well—as I once heard it described, "It's not information overload that's the problem, it's organizational underload!"

WHY DO WE NEED NOURISHING SLEEP?

Well, I know you don't want me to bore you with stuff you already know, so I am going to reiterate the boring stuff super quickly and then give you some information that you might not be familiar with!

We need good sleep so we feel good, so we have energy, so our memories work, and so our brains have time to clean themselves. Animal studies show that our brains shrink so the cerebrospinal fluid, which surrounds our brains, can increase in volume and "wash" our brains as we sleep. How cool is that?

The different stages of sleep have different functions. The first two are relatively brief, but are characterized by changes in brain waves. The third stage of sleep, called deep sleep, is vital for improving immune function, repairing muscles and tissues, and building energy for the next day. REM—rapid eye movement—is the stage we go into after the deep sleep and is when we dream, consolidate memories, processes learning. These are all vital for us to live healthy, productive lives, *n'est-ce pas?*

Not getting enough sleep will make you crazy—both in your mind and in your body.

If you've ever had children, or been a caregiver of a new puppy or someone who requires constant monitoring during the night, you know how hard it can be to be woken every few hours. In fact, it's a known torture technique—not allowing people to sleep. Variable sleep schedules also create trauma.

Disordered sleep kills your sex drive, creates disorientation, leads to depression, ages your skin, and can increase weight gain. It has been linked to heart disease, heart attack, heart failure, irregular heartbeat, high blood pressure, stroke, and diabetes.[32]

---

32 Institute of Medicine. Sleep Disorders and Sleep Deprivation: An Unmet Public Health Problem. Washington, DC: The National Academies Press; 2006.

Phew—who needs that? I like having a healthy sex drive, focus, a more stabilized mood, young-looking skin, and physical health—and I am guessing you do, too!

HOW TO SLEEP WELL

Date yourself. Seriously—set yourself up on a sexy, slow, loving evening date. If you're Mama Gena and you have a vagina, she'd tell you to pussify your life. Here's my version for everyone!

Go to bed early and get up early—the early part of the day is the best for getting things done. You may want to roll over and take advantage of your lover's morning glory (if so, set your alarm a little earlier!), but getting up and getting on with your day is what makes the difference between the successful and the not-quite-there-yets, and it will make you feel GREAT!

"A person who has not done one-half his day's work by ten o'clock, runs a chance of leaving the other half undone."

—Emily Brontë

Remember the Ayurvedic Clock from Chapter 2? Pay attention to it (Figure 3). The most important piece of information I want you to absorb here is the TIME. At 10:00 p.m., that Pitta fire starts to burn again. Now, if you're up and about and not paying attention to your mind-body, you will find that you get that second wind, that "night owl" energy, which burns from 10:00 p.m.–2:00 a.m. This Pitta fire is much better directed at your mind-body while you're asleep. This energy is great for allowing your digestive system to process and detox (the liver starts its important work at this time) and your brain to process and digest the experiences of the day.

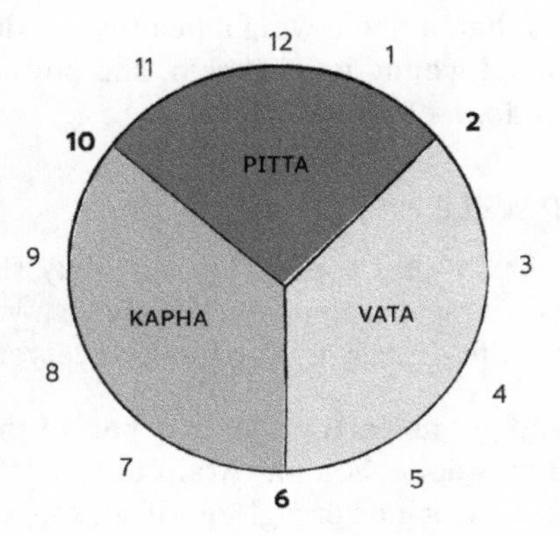

Figure 3. The Ayurvedic Clock

So, how are you going to accomplish this? By organizing your food, your lights, and your senses, that's how, my friends! Read on…

FOOD

Look at the Ayurvedic Clock. Have a light, plant-based supper by 6:00 p.m. A light, plant-based supper? Head to the back of my book for a recipe book written by *moi*, which has simple, tasty, plant-based menu items you can incorporate into your life. Did you know that supper is an English word and it means supplemental, i.e. this should not be your main meal of the day? It's hard to go to sleep if you have a huge steak at 8:00 p.m. Why? Well, you're eating in the Kapha time (6:00–10:00 p.m.), which is a slow, heavy, watery-earthy energy, and it does not have the digestive fire to help you deal with lots of animal protein.

Essentially, you want to digest your food before you lie down. Gravity is no longer on your side when you're in bed, and let's face it, sex is much more fun if you don't have a belly full of food, am I right? Also, you will find it much easier to wake in the morning when your sleep is not disturbed by a night of digestive struggles.

LIGHTS

Look at the Ayurvedic Clock: 6:00–10:00 p.m. is Kapha time. This is why, if you allow your body to follow its natural rhythms and don't disrupt them, you will start to feel that heavy, slow, earthy-watery relaxation. There are two simple ways to help bolster this shift and take advantage of Kapha time.

1. **Dim the lights and cut out the technology after supper.** Whaaat…? But but but, everyone's on Bumble in the evenings and I have a deadline and I want to read my book on my iPad and and and—listen, do you want good sleep? Then try this! The blue light in TV, computer, tablet, and smartphone screens stops the production of melatonin. Melatonin is a hormone that is responsible for your sleep-wake cycle or circadian rhythm.

2. **Boost your sleep-wake cycle by getting some sunshine during the day.** Also, increase melatonin production by being outside as dusk is falling. Then come home and sit down by a dimmed light and read a book (the most successful entrepreneurs in the world supposedly spend five hours a week doing things they enjoy that are not related to their businesses—you can, too), knit a scarf, talk to someone, have a bath, have sex (with yourself or someone else!), batch cook some food for the week, play a game, or read a book in another language. Light some candles, set the diffuser, or get some incense going. Date yourself. Seriously, set the scene—don't just put the effort in for other people!

SOOTHE YOUR SENSES

Create an evening routine, e.g., turn off all technology, take a walk, soak in a warm bath, massage your feet, read a book. Do the same order of habits every night for a week and write down how you feel.

Soothe your senses. Because it's Kapha time (6:00–10:00 p.m.), you should work on pacifying and relaxing. Again, this goes back to dating yourself. What does that look like? Reduce the cortisol and pacify your nervous system—a.k.a. chill the f*ck out.

Remember that dating yourself means no arguments, and no stressful interactions about watching violent TV shows or who unloaded the dishwasher. Got it? Turn on some chill music (think Al Green's "Tired of Being Alone" rather than Marilyn Manson's "The Fight Song") and begin.

Here's how:

1. **BATHE:** If you have a bathtub, go fill it with warm water and make it sumptuous. Maybe light a candle or get the infuser with essential oils going. Add any or all of the

following: Epsom salts (1 cup), Himalayan salts (1 cup), or baking soda (1 cup). All are detoxifying and good for the skin. By the way, detoxing means drawing the gunk (or *ama* as we call it in Ayurveda) out of your body.

2. **USE OILS:** Add some relaxing essential oils, such as lavender, vanilla, and cedarwood (avoid citrus, which is creative-inspiring, and peppermint and eucalyptus, which are stimulating). It's good to get your oils from reputable places, where they are not buggering with the environment or people or using yucky and harsh chemicals. I like Doterra and NOW brands for my essential oils. Play with the amounts. Try adding five to ten drops of oil to your bath and see how that works. Remember essential oils are powerful—have fun! Or just massage your hands and feet with some coconut oil to signal that their work is done for the day.

3. **ABHYANGA:** Self-massage with warmed oil. Now, while the bath is running, and the room is getting warm and steamy, give yourself some self-love. Now now, not that kind (at least not while I'm watching). Er, where was I? With some coconut, sesame, or almond oil (food grade is more highly purified than body grade), massage your body from head to toe. Rub the oil into your feet and ankles, up and down your legs, around your knee joints, your buttocks and lower belly, all over your abdomen and lower back, making a U shape from right hip to right lower rib to left lower rib, to left hip, which follows your digestion along the ascending, transverse and descending colon. Smooth the oil into your hands, into your nails, and into your wrists and arms. Slather your boobs/moobs and upper chest. Drizzle more oil into your hands, warm it up, and rub in circles on your shoulders and between the shoulder blades if you can get there. Then put some oil onto your hair and face, massaging deeply and lovingly. Again, this is about dating yourself.

   Abhyanga is a practice for soothing the mind as much as it for soothing the body. It's also really fricking important to face your whole body and pay attention to it on a regular basis, by loving your body and connecting your mind to it. Tell your lovely, cellulite-y thighs, "thank you for helping me run up the stairs" (cellulite is subcutaneous fat, so when you start to cut out the fat of animal products and massage your thighs, your cellulite will decrease. Just sayin').

Why use oil? Well, your skin is your body's largest organ. Whatever you put on your skin goes straight into your blood, so you want to be super mindful about that. You know that nasty smell of rancid oil? Well, the processes that occur to make a lotion essentially prevent us from knowing whether the oil has gone rancid. Yuck. Also, many lotions have parabens, phthalates, and lots of other nasty stuff, which can disrupt our hormones or have other ingredients like dimethicone, which stops the skin from breathing. The skin is designed to release as well as absorb.

Now, get into that tub and soak. The warm water will open the pores and the oil will soak into a deeper layer of the dermis. Whatever doesn't soak in can be gently wiped off. It really is the best thing ever for your body and your mind. As your bath is draining, squirt some dish detergent (earth-friendly/non-toxic) down the drain so your pipes don't block up (similar principle to the kitchen sink and the fats that you cook with).

4. **MAKE SOME WARM HERBAL TEA OR GOLDEN MILK.** Or, if you tend to wake at 2:00 a.m. and can't fall back to sleep again, try eating a slow-release carb like a sweet potato at supper and then make some Golden Milk. Put a cup of milk (nut or cow's) in a saucepan. Add sedative spices, such as a half teaspoon of vanilla and one-quarter to one-half teaspoon of each of the following: cardamom, cinnamon, nutmeg, and turmeric.

   Unsweetened nut milks can taste, well, un-sweet if you've been consuming cow's milk for years. So, perhaps stir in a half teaspoon of maple syrup or a tad of stevia, and then reduce the amount over a week or so until you've weaned yourself. Just be aware that you will want to stir or whisk it up to blend it. Now, from reading this, you're probably getting a sense about me that I am usually all about the hacks to make this stuff doable. But sometimes it's worth taking the time. What I have found, with making Kombucha and Golden Milk for a while, is that you get much better flavors via the infusion with roots, sticks, barks, etc., rather than the ground varieties. So, I prefer using two to three cardamom pods, lightly crushed, one-eighth stick of cinnamon, a chunk of nutmeg and a little turmeric root and vanilla essence, which infuses the milk with yummy flavors. Remember, the key is to reduce the barrier to action, so do what works (when I travel, I make a blended shaker of the spices to stir into whatever I can get my hands on, milk or herbal-tea-wise).

5. **FINALLY...** Only sleep and have sex in your bed. Do not associate your bed with anything else—no working, reading, eating, watching TV, or texting. Your bed is a place of respite, nourishment, and sensuality. This is a basic sleep hygiene notion that is SUPER important to initiate. Otherwise, the space is associated with energy that is not conducive to sleep and post-orgasmic mmmmmmm-ness. Got it?

"Sleep is God. Go worship."

—Jim Butcher

TO NAP OR NOT TO NAP?

I am a fan of naps. They take a while to get good at for some of us, but for others it comes more easily! I've read all sorts of tricks over the years, as a new mom, a working mom (who would reliably fall asleep onto my computer keyboard after lunch when pregnant!), and as a mom of three!

Here's how to nap in a healthy way—with a couple of caveats, which are similar to what some of the pediatric sleep experts will tell you.

1. Take a short "power nap," not a long one. If you have one of those ninety-minute naps, it's really hard to wake up and get out of the foggy feeling. Caveat: if you have had a very disturbed night, then try to make up to a total of eight hours' sleep within the twenty-four-hour cycle by taking these longer naps. Get some of that deep sleep and maybe some REM sleep (about ninety minutes after falling asleep) by sleeping for a longer time!

2. Don't nap after 3:00 p.m., because otherwise you will find it hard to go to sleep early and you will perpetuate the cycle again.

3. Use ear plugs and an eye mask to help you quiet your nervous system.

SUMMARY

> Observe the connection—do you sleep better when...? Do you sleep worse when...?

Okay, so by now you should totally be buying into the fact that sleep is a major part of being healthy. I've given you lots of good info. Now you need to make this stuff happen for you. So, remember, start with a light supplemental meal in the evening (a.k.a. supper). Use lights to your advantage, and then work to soothe your senses.

WHAT DO YOU NEED TO UNLEASH YOUR POWER?

Here are some easy items to add to your grocery list, plus a quick, bossy, hell-no-get-rid-of-that list, which you can do to make the evening go more smoothly.

Evening Routine

- Eat a small, plant-based meal by 6:00 p.m. to digest before bed.
- Stay away from melatonin-reducing activities (TV, phone, computer) and go for an evening walk.
- Avoid arguments and stressful activities in the evening.

REMEMBER: Your day starts the night before!

Bathing Goodies

- Epsom salts
- Himalayan salts
- Baking soda
- Soothing essential oils (e.g., vanilla, lavender, cedarwood)

Abhyanga/Self-Massage Goodies

- Almond or sesame oil (base oil)
- Reusable glass/pottery soap dispenser (optional, but makes oil dispensing super easy)
- Essential oils to add to base oil so you smell yummy (I like neroli, patchouli—yes, I'm a hippie chick—rose, vanilla, lavender, jasmine, and ylang ylang)

Evening Drink Goodies

- Chamomile, lavender, and/or valerian teas
- Milk and sedative spices (cinnamon, vanilla, cardamom, turmeric, nutmeg)

Bedroom Items

- Oil diffuser (get one of those safe, switch-off varieties, with no BPA)
- Eye mask
- Ear plugs

REMEMBER: Get rid of TV's, radios, and phones in your bedroom!

THE MAIN MESSAGE

Eat early, pacify your senses, and go to bed by 10:00 p.m.

How?

Here are a few basic and doable tips for you to start making good sleep a reality. Don't overthink it. Just try it and start to pay attention to how you feel. You're ready to unleash your power, aren't you?!

1. *Reverse engineer your evening*. Want to be asleep by 10:00 p.m.? Plan to get into bed at 9:50 p.m. Drink tea or milk, massage feet, read a book from 9:30–9:50 p.m. Use the bathroom from 9:10–9:30 p.m. Shut up house, let out your dog or lover, turn off lights, etc., by 9:05 p.m.

2. *Put your bed and wake times in your smartphone calendar, on repeat*. Write Post-it notes and post around the house. Tell your housemates and/or friends who you socialize with in the evenings that you want to be asleep by 10:00 p.m.—set up your accountability partners.

3. *Track your progress* and be your own habit scientist. Use an app or an activity tracker and/or write down what you do. Take note of how you feel when you do/don't follow these suggestions.

TIME OUT

You know what? The universal truth with all this is that most of us will resist these regimens! We resist because we're bored, because we feel rebellious, because we're watching an awesome show on Netflix, because we're having a super hot texting conversation with someone from Bumble, we resist…

STOP, Hammer Time.

TIP 1: IMPERFECT ACTION

What would happen if you just tried it for four days this week? Notice how you feel when you go to bed at the same time every day, get up at the same time every day, and create a soothing evening routine. It doesn't have to be perfect. I am a huge fan of "imperfect action." Don't obsess with getting it (perfectly) right, obsess with getting it done. JUST DO IT (I'm starting to feel like a Nike commercial). The successful people in the world are the ones who go for it before they feel ready, before they have all the pieces in place, or when they still feel like it might not work. The key is consistency of the habit.

TIP 2: FALLING OFF THE WAGON CAN HELP YOU IDENTIFY YOUR KEYSTONE HABIT

The real key is to get back on again and get to it, without a lot of drama. Use the time off to notice (recall in Chapter 3, I said I want you to be a habit scientist, too—someone who really starts to pay attention to yourself) whether you kept any of the habits going and, if so, how did they make you feel? Many of us have one clear "keystone habit" (as Charles Duhigg calls it)—the habit that, if we can nail it down, will create this ripple effect of the other habits just slotting into place, like the keystone of an arch.

For example, your keystone habit might be sleep. If you sleep seven hours every day, then you make better food choices, your energy is better maintained, you feel better and you exercise. When I let my keystone habit slide, then the proverbial shit hits the fan. So, if I'm traveling or out of my usual schedule, I know that I have to work hard to keep that habit in check, and then I'm golden. Got it?

1. Put these practices into place. Commit to them.
2. Once these habits are really established, it's time to start playing the role of the habit scientist. Try dropping a habit.

I like to do this when I travel or when I have guests, which naturally disrupt my regular schedule.

3. Notice whether that habit sets you off course, or whether you are still able to do the other habits that you need to do. For example, I can go two or three days without meditating, and then my reaction time to irritation shortens significantly. I can go three or four days without eating a plant-based diet, and then my gut starts to complain, I feel lethargic and slow. This is really about KNOWING THYSELF. Got it?

4. **BONUS:** While you're on vacation or during the weekends, try to avoid the stay-up-really-late, sleep-really-late model, and be your own habit scientist by trying to figure out your natural sleep-wake cycle without adding socializing, seeing live music, parties, and getting up for work into the mix! I've discovered that my natural cycle is 9:30 p.m. to 4:30 a.m. I regularly and reliably wake up at 4:30 a.m. when I have gone to sleep early enough (or at least in that ballpark), and if I really pay attention to my body and its signals, I start to fall asleep sometime after 9:00 p.m. If I let myself stay up, then the fire of Pitta starts to ignite and rather than working on digesting toxins and allowing my liver to work its magic and detox my brain, it keeps me awake and fired up and I find myself wanting to start projects!

WHY? If you sleep better, you'll eat better, work better, and have more energy, and love your life.

HASHTAGS #Energy #Melatonin #BoostMelatonin #CircadianRhythms #GoOutsideAtDusk #TakeANap #WakeUpEarly #GoToBedBy10 #SleepHygiene #SleepMakesYouSexy #SleepMakesYouHealthy #SleepMakesYouHorny #SleepIsGoodForYou #Abhyanga #Massage #Detox #OilYourself #DateYourself #TakeBaths #EssentialOils #GoldenMilk

CHAPTER 5
EATING

This is the section where I tell you not to eat doughnuts and to eat lots of kale. Boring, I know. But it's one of the linchpins of having a free and prosperous life. Your body houses your mind, so don't get obsessed, but do take good care of it. Mmmkay?!

"One cannot think well, love well, sleep well, if one has not dined well."

—Virginia Woolf, *A Room of One's Own*

SCARCITY AND NOURISHMENT

Often when I work with my clients, and we start to talk about making shifts in diet, they move into scarcity mode. The common perception is that life will be bland, flavorless, and without color or texture if you start to remove or cut down on the amount of steak, cheese, doughnuts, and eggs that you consume. People often feel that this is NOT a great habit to develop! To which I say: This is bollocks.

Food, along with everything else that you consume through your senses, is about nourishment. Nourishment is the opposite of deprivation—it's about growth, health, and good condition! It's about what you can do that works well for you. So, instead of seeing a shift to more vegetables and less animal products, from processed to whole foods, from internationally shipped food to local food, from loads of simple carbs to more complex carbs, as a negative, see it as a positive.

To nourish means to "supply with what is necessary for life, health, and growth," and "to cherish, foster, keep alive," as well as "to strengthen, build up, or promote." This is positive, don't you think? This is what we want from our food, and our lives in general, in my opinion.

One way to really start to make that shift is to work on the mindfulness aspect of food and eating. Yada yada yada, I hear you say. But, stay with me!

HOW TO BE MINDFUL ABOUT FOOD AND EATING

Just take a moment, and if you're at home, go and open your fridge (or just visualize it for a moment). Start by thinking about where your food came from—the journey it's taken to get to you.

Let's start here: the seeds are grown and cultivated, then packaged and transported to a farmer. The farmer plants them and then tends to their growth. Then the plant (e.g., carrots) is cultivated by the farmer, transported to a site where it's processed (e.g., your baby carrots are actually bigger "imperfect" carrots, which are sliced into approximately two-inch lengths, smoothed and rounded on the ends, polished, and, finally, washed in chlorinated water to remove bacteria). Then the carrots are packaged and transported to the grocery store. I didn't even mention the trading and storage along each step which, for fresh produce, means refrigeration.

So, when you sit down and look at your plate of a variety of foods, pause and consider their journey—all the people who have made it possible for you to have a plate of such color, vibrancy, and texture. Then, really look at the food—the colors, the shapes, the smells. Imagine the textures on your tongue, such as whether the food is soft, crunchy, or chewy (and if your plate is all one color—think again, my friend, about "eating the rainbow"). This practice makes me feel like I am nourishing myself—cherishing the experience and the abundance.

If you're eating with others, look at each of them, and thank them for taking the time to eat with you and their role. For example, maybe your son laid the table, your daughter made the salad, or your spouse lit some candles; or perhaps your friend suggested you meet and try out the new taco restaurant. Perhaps ask them what they are grateful for today as you sit together. This fosters their role in getting the food and the experience together and their importance to you.

Doesn't this feel less scarce and more abundant now?

SOME OF THE SCIENCE BEHIND THE STANDARD AMERICAN DIET (SAD)

You want to live long and prosper, right? You want to avoid heart disease and diabetes and maybe even dementia? You want to be able to get and maintain an erection if you're an *hombre* and be fully (naturally) lubed up if you're a *chica*, don'tcha? I know I do, ladies and gents! Simply put, you're more likely to do these things if you eat more plants and keep away from the Standard American Diet (SAD). Need some inspiration? Flip to the back of this book and you'll find a great plant-based recipe book for every meal, including some sweet treats.

I can tell you that I don't want to get stuck with having to eat the SAD. You know why? Because it's got an AWFUL name and it's pretty crappy, too (are you eating SAD?!). The SAD includes predominantly refined grains and sugars (real and fake), which have an appalling effect on your body, and huge amounts of fat, dairy, and meat, which are totally unnecessary. According to the US Department of Agriculture in 2016, one-third of Americans' daily calories come from added fats and sweeteners, which includes anything with "-ose" in it, whereas only about 8 percent of our calories come from fruits and vegetables.[33]

Part of the problem stems from the residual misinformation around fats that started that whole fat-free movement in the 1980s. Low-fat diets were pushed for people with heart disease, yet physicians and then the federal government pushed low-fat diets for all. The problem was that the food tasted bland, so sugar was added. And people got fatter—not thinner—as they dove into all those disgusting fat-free alternatives to real food, such as fat-free cookies and ice cream, salad dressings, and dairy products. In the words of Rory Freedman, a.k.a. the Skinny Bitch, "Whenever you see the words 'fat-free' or 'low-fat,' think of the words 'chemical shit storm.' "[34]

Fats, like many other things in life, are not created equal. I am not going to give you a big science lesson here, because you'll get bored and not continue reading, but I just want you to grasp the basics. Sound good?

The fat in an avocado or in nuts is good for you, in small amounts (unless you're in the advanced stages of coronary heart disease), but the hydrogenated fat (fat with added hydrogen so that it stays solid at room temperature) that is used to fry your French fries or doughnut, and make your crackers, cookies, and store-bought mayonnaise, is not good for you. When I say "not good for you," I mean that they mess up your system! How? It's because of the way these fats are manufactured.

33 "A Look at Calorie Sources in the American Diet." USDA Economic Research Service. http://www.ers.usda.gov/amber-waves/2016/december/a-look-at-calorie-sources-in-the-american-diet
34 Freedman, Rory. *Skinny Bitch*. Philadelphia, PA: Running Press, 2005.

Hydrogenated fats have a chemical structure that the body finds hard to either metabolize (and then excrete) or to lay down in your cells. Thus, they circulate in your blood, elevating your LDL cholesterol (bad cholesterol) and contributing to blocking your arteries and possibly cancer formation.

Your body is a magnificent, remarkable feat of engineering. Your blood moves around your body via a group of vessels: the veins (which carry deoxygenated blood to the heart) and the arteries (which carry oxygenated blood away from the heart). When you consume vast quantities of fat, the inner layer of your arteries starts to break down, and like a cut on your finger, will develop a scab. The problem is that on the inside of your artery, there is limited capacity for blood flow, so if that scab breaks free, it will block up your artery and it may burst. Can we say heart attack or stroke? Not good.

The inner lining of the artery—the endothelium—makes nitric oxide. This is a gas that is vital to the correct functioning of your arteries because it allows for the expansion of the artery, which allows more blood to flow. This information sure comes in handy if you need to run away from your mother-in-law—I mean, that woolly mammoth in the forest!

Listen up, guys. Do you know where else it's important to get a good expansion of the arteries and compression of the veins? Your penis! You can't maintain an erection if you don't produce nitric oxide. So, if you (or your lover) are having problems in your pants, maybe it's time to look to your diet (if emotional and other physical causes have been ruled out).

Gals, too. Don't think you're not affected, missy! When a woman is primed and ready for sex, her clitoris becomes engorged (blood to the capillaries) and her vagina gets well-lubricated (some juicy lady chemicals plus plasma from the blood). Jane Esselstyn, an amazing plant-based nurse, educator, and writer of the recipe sections of her brother Rip's *New York Times* bestselling recipe books, told me when I interviewed her that, in her family, they say, "Plant-strong: Pant-strong."

'Nuff said.

The fat also collects around your internal organs—your liver, kidneys, and heart—which makes it hard for them to function correctly. I could continue on this tangent, but you all get the picture and know this stuff. What you may be less aware of is how what you eat can change your hormones, your brain, and how you feel. This field of study, called psychoneuroendocrinology, is seriously sexy. It looks at how your hormones (endocrine system) affect your moods (neuropsychology). Cool, right? I know, I am such a NERD and I love it.

THE STANDARD AMERICAN DIET (SAD) WILL LITERALLY MAKE YOU SAD

So, back to what you're eating. I sure as hell don't want to eat SAD because it will actually make me sad, as well as de-lube my vagina! Wanna know why it'll make me sad? It all comes down to the gut.

The gut is amazing. Like totes amazing. It is our second brain. Also, the journey our food takes is long—super long. The alimentary canal is a thing of much beauty and general awesomeness, don't you think? You chew on an apple and your teeth cut it up. Then they shred it, and the saliva starts the initial process of digestion. The stomach acid slowly starts its work and then after a few hours, the apple's journey into the small intestine starts, which takes from a few hours up to a few days, and then it moves on to the large intestine or colon. Generally, there is a huge variability in digestion from person to person. Those peeps who have more Pitta (fire and water) in their systems tend to move food faster than those with Vata (air and space) or Kapha (earth and water). Their digestive fire, or *agni,* is stronger.

Not blessed with a cast-iron Pitta digestive tract? Fear not, my friend. Those of us who eat more plants and less starch and heavy animal products also move food through our systems way faster (you may poop once a day or multiple times a day when you're eating a plant-based diet).

There are so many nerves in our stomach and intestines, and there is more serotonin in our guts than in our brains. Serotonin, you know, is that pesky hormone that contributes to our happiness—and is impacted by many prescribed antidepressants, like Prozac, which are called selective serotonin reuptake inhibitors (SSRI).

So, what you guzzle can totally affect how you feel, psychologically and emotionally—not just whether you have a muffin top, or whether your inner thighs chafe, or how your sense of heaviness and sluggishness might make you choose the baggy outfits hiding in the back of your closet (the ones that would be best relegated for donation or torn up and used to clean your house).

All said, one easy thing you can do while at the grocery store is not only load up on plants—a.k.a. whole foods—but also take a closer look at the other things you're putting into your cart. Meaning: investigate the ingredients in them.

READ INGREDIENTS LABELS!

The key is to get good at reading ingredients labels and aim to avoid consuming foods with lots of added crapola. Some of the easiest switches to make include salad dressings (use olive oil and vinegar or lemon juice, or hummus and balsamic vinegar and mustard), sauces (try ketchup—look for brands with no high-fructose corn syrup), and drinks (switch to sparkling water with a wedge of lime or cucumber, or a sprig of mint, rather than guzzling toxic diet sodas).

WHAT'S HAPPENED TO OUR DIET?

Our diet has fundamentally shifted in four main ways.

1. ***Access to food.*** We can get it anywhere, anyhow: drive-thru, twenty-four hours a day, when buying gas for our car or ibuprofen, or from a vending machine in the library.

2. ***When we eat.*** We tend to eat our large main meals later in the day, which—if you remember the Ayurvedic Clock—is NOT good (Kapha time is 6:00–10:00 p.m., earth-water time, which means low digestive fire, AKA bile.).

3. ***Portion size***. A McDonald's serving of French fries in 1955 was 2.4 ounces, whereas today it's 5.9 ounces (this serves up a whopping 510 calories, which is over one-quarter of an adult's recommended daily allowance of calories).[35] It's no surprise, then, that the average weight of a man in the United States went from 166 lbs. in 1960 to 191 lbs. in 2002, and a woman's average weight has increased from 140 lbs. to 166 lbs.[36] Yes, we have gained about an inch in height, but that doesn't explain all of the weight gain. Bear in mind that the average height of a woman in the United States today is 5'3", which means her Body Mass Index (BMI) is 29.4 (obese is 29.5). For men, the average height is 5'7", making his BMI in the obese range at 30.5. Curiosity led me to calculate mine at the time of writing this in 2017:

---

35 Bratskeir, Kate. "McDonald's Portion Sizes Have Drastically Changed Since 1955." *The Huffington Post.* September 10, 2015. http://www.huffingtonpost.com/entry/mcdonalds-portion-sizes-over-the-years_us_55f171b4e4b002d5c0782ee8.
36 "Americans Slightly Taller, Much Heavier than 40 Years Ago." The Centers for Disease Control and Prevention. October 27, 2004. https://www.cdc.gov/media/pressrel/r041027.htm.

I'm 5'10" and 150 lbs. so my BMI is 21.5, which is in the healthy range of 18.5 to 24.9. Oh, and don't forget, if you make your own BMI calculations, that muscle weighs more than fat so, please don't get hung up on this damn number!!

4. ***The balance of different food groups.*** Not all calories are created equal. That's also at issue here—eating 510 calories of French fries is a totally different experience for your body than eating 510 calories of plants. We tend to get most of our calories from the wrong food groups (fat, sugar, processed carbs, etc.), rather than plants. We eat way more processed food now than we used to, and, to keep the food stable, lots of salt and other stabilizers are added which do not benefit our bodies or minds!

Trying to drop a few extra pounds? Keep a food diary. Note down what you ate and when.

Rant alert ahead: Calorie-counting and dieting is BS. Who wants to spend their lives being obsessed with whether this or that has more calories and what you're "allowed" to eat? Ugh. Dieting is a disaster—people put on more weight after they've come off the diet than they lost in the first place (may even be as many as 80 percent). Weight gain may be due to hormonal disruptions brought on by the dieting, which affect hunger levels for a long time post-diet.

Numerous studies, from twin studies (rules out genes) to long-term studies (follows long-term behaviors) to meta-analyses (lots of studies compared), all show that being on a diet increases your likelihood of weight gain later on and that dieting DOES NOT WORK.[37] [38] [39] You want to cultivate a way of life where food is a nourishing, joyful, fun, yummy, and feel-good part of your day, rather than a negative, obsessive, and restricted field of explosive decisions and stress.

"The best diet is the one you don't know you're on."

—Brian Wansink, PhD

So, just be super mindful about what you choose to put in your gob. Got it? Read the ingredients labels while you shop, buy food that is whole/organic/non-GMO/local, is from the perimeter of the store, think twice before putting certain items in your cart, learn to appreciate where your food comes from, and plan your meals

37 Wolpert, Stuart. "Dieting Does Not Work, UCLA Researchers Report." UCLA Newsroom. April 03, 2007. http://newsroom.ucla.edu/releases/Dieting-Does-Not-Work-UCLA-Researchers-7832.
38 Pietiläinen, KH. "Does Dieting Make You Fat? A Twin Study." *International Journal of Obesity* 36 (2012):456–64.
39 MacMillan, Amanda. "After Dieting, Hormone Changes May Fuel Weight Regain." CNN. October 26, 2011. http://www.cnn.com/2011/10/26/health/post-diet-weight-regain.

around and eat predominantly whole foods. Think eggplant and asparagus, not steak and/or pasta. Our bodies are designed to crave fat and sugar, but we eat way more of those than we need. Not that long ago, we used to grow our own food, or at least buy it from or trade it with our neighbors. To find a way back into that super powerful person you know you're supposed to be, start to focus on developing these great habits!

SAY NO TO PROCESSED FOOD AND ARTIFICIAL SWEETENERS

Did you know that eating sugar-heavy foods, such as candy and ice cream, at bedtime can create an insulin spike about four hours later, which can WAKE YOU UP!!? Replace with a banana or some dates, or a cup of herbal tea.

Now, the majority of us in the west don't know where our food has come from, what journey it has taken to get here, how many hands it has touched, or what's in it! So, there is a tendency to eat way more processed food than we should, which has too much salt, sugar, and fat and not enough fiber, minerals, and vitamins. Also, it feels convenient too, doesn't it? Having food that lasts a long time in the freezer or cupboard.

Some people try to skirt the issue of sugar by consuming artificial sweeteners. Firstly, you will start to notice that these just taste gross when you "clean up" your taste buds. I know, it sounds weird, but the first time I did a cleanse, I could not believe how sweet white rice and cow's milk were—the sweetness just popped! And I wasn't tasting that when I was eating more sugar. It's all about the baseline. But, more important, the research points out that artificial sweeteners are dodgy AF.

Canadian researchers, published in the journal PLOS One, followed fifteen hundred people for an average of a decade, half of whom consumed artificial sweeteners and half of whom didn't.[40] Those who used artificial sweeteners had a 53 percent higher amount of abdominal obesity, 2.6-cm larger waist circumference, and much larger body mass indices. Seems weird, right? But artificial sweeteners reduce your glucose tolerance and alter the function and composition of intestinal microbiota. That spells long-term disaster, my friends! Eat a date (the edible sweet fruit, not a potential lover) instead.

40 Chia, Chee W., et al. "Chronic Low-Calorie Sweetener Use and Risk of Abdominal Obesity among Older Adults: A Cohort Study." PLoS ONE 11 (2016). http://journals.plos.org/plosone/article?id=10.1371/journal.pone.0167241.

"When your body is hungry it wants nutrients, not calories."

—nourishingourchildren.org

FOOD IS NOT RATIONAL

I know you're smart, otherwise you wouldn't be reading this book, so I am not going to bore you with stuff you already know. What I will do is give you some tips that you may not have thought about, which are grounded in science.

But first, food is complicated by our childhoods, our memories, and our traditions. Many of us grow up with strongly-rooted experiences of food, people, and events. For example, our grandmother's apple pie for Sunday lunch, our mother's shepherd's pie for Thursday supper, or a large bowl of chicken noodle soup when we were sick.

Best source of probiotics? Ones that start at the beginning of your alimentary canal, a.k.a. your mouth, like kombucha, sauerkraut and kimchi.

We associate major celebrations with food—the turkey at Thanksgiving (or Christmas if you're a Brit), Victoria sponge cake with strawberry jam between the layers for your birthday, and caramel apples and apple fritters in the fall. Maybe we have memories of traveling to different places, which are deeply connected with food. I will forever associate my trips to South Carolina as a young adult from London with buttery homemade biscuits, eaten with lashings of melting butter and honey, served on the side of slightly spiced black beans and fried shrimp. Yum, yum, yum. I'd not eaten that combination of salty and sweet growing up.

So, what I am saying is, do not discount the power of memories, traditions, and people, and how these things connect to food.

"Food is not rational. Food is culture, habit, craving and identity."

—Jonathan Safran Foer

Approximately 35 percent of Americans are obese and approximately 30 percent of Brits and Kiwis are obese.[41] Abdominal obesity is a major issue here—that's the fat around the midsection, which is closely connected with cardiovascular disease, type 2 diabetes,

41 "Obesity Statistics." New Zealand Ministry of Health. January 24, 2017. http://www.health.govt.nz/nz-health-statistics/health-statistics-and-data-sets/obesity-statistics.

and cancer. That's a distinction between carrying fat around your hips and thighs and your midsection.

Some scientists have created charts to measure the hip-to-waist ratio as a marker for this (which is different for men and women). The issue is that the fat that is stored around your midsection is stored around your major organs. This fat is more metabolically active, which means it's releasing more fatty acids, inflammatory agents, and hormones.

What does that mean? It means that we're killing ourselves with food. That's scary. It's even been predicted that our children will die before us, due to the crud they put in their mouths.[42] How insane is that? Now, this is not a book about civil rights and access to food, which I know has a huge impact on this issue, but let me start by telling you that what you put in your mouth can fundamentally affect who you are. I don't think enough of us are paying attention to this.

There is also a lot of political (geopolitical?) energy around food pyramids and the stats on how much dairy and grains we should be eating each day, which are perhaps not always grounded in science. For example, consider the bogus notion that milk makes our bones strong, pushed by the dairy farmers and that "Got Milk?" campaign. In fact, studies show higher levels of milk consumption may have a deleterious effect on health, e.g., increased rate of death; increased hip fractures; increased risk of ovarian, breast and prostate cancers; disruption of hormone production in men (reduced testosterone); and the disruption of sexual maturation in pre-pubertal children.[43 44 45] Let's face it, cow's milk is designed for the digestive tract of a calf, which has four stomachs!

Plants such as kale, broccoli, collard greens, and spinach have calcium and are all great alternatives to milk and other dairy products.

Many of us maintain deeply held beliefs based on what our mothers or grandfathers told us. For example, you must eat animal protein with every meal if you want to be strong! Remember to challenge these beliefs. Protein can be found in plants, which many people do not believe, even though the science says otherwise.

---

42 [42] Olshansky, S. Jay et al. "A Potential Decline in Life Expectancy in the United States in the 21st Century." *The New England Journal of Medicine* 352 (2005):1138–1145. http://www.nejm.org/doi/full/10.1056/NEJMsr043743#t=article.
43 Michaelsson, Karl, et al. "Milk Intake and Risk of Mortality and Fractures in Women and Men: Cohort Studies." *BMJ* 349 (2014). http://www.bmj.com/content/349/bmj.g6015.
44 "Calcium: What's Best for Your Bones and Health?" Harvard T.H. Chan School of Public Health. https://www.hsph.harvard.edu/nutritionsource/calcium-full-story.
45 Thomssen, Christoph, ed. "Consumption of Cow's Milk and Possible Risk of Breast Cancer," *Breast Care* 5 (2010):44–46. https://www.ncbi.nlm.nih.gov/pmc/articles/PMC3357167.

For most of us, we only require 10 percent of our calories from protein (unless we're hardcore athletes, pregnant, or nursing—in which case, eat more protein-rich plants, duh!), which is totally doable by eating plants. Have you seen some of the extraordinary super-athletes out there who are plant-based? Google the images and stories of these people if you don't know about them: Serena Williams (winner of 23 tennis titles), David Haye (six-packed boxer), David Meyer (winner of seventeen national and international martial arts championships), Tia Blanco (multi-award-winning surfer), Heather Mills (skier), Rich Roll (Ultraman 2009, from an overweight couch potato to a plant-powered middle-aged Superman!).

The suggested amount of protein is approximately 45 grams (g) per day for a woman and 56 g per day for a man, or 0.36 g per pound (lb.) you weigh. So, take me as an example. Right now, I am 150 lbs., so in this model, I require 0.36 x 150 = 54 g/day.

This is easy to do with plants. When I start my day with a cup of cooked oatmeal and two tablespoons of chia seeds (plus a half cup of blueberries and a banana) I've got 12.8 g of protein under my belt already (oatmeal, 6 g; chia seeds, 5 g; blueberries, 0.5 g; and banana, 1.3 g)! Then, for example, I can eat ½ cup of lentils (9 g), a cup of peas (8 g), a half cup of quinoa (4 g), and two cups of spinach (10 g) for lunch, which totals 31 g—and I've hit 43.8 g of my daily required 54 g consumed by 1:00 p.m. You dig?

The idea of what we eat and when—the history of food and eating—is so interesting. If you've ever traveled, or even just spent time with other families, you know that it's not unusual to eat different foods at different times of the day. For example, when I was in Thailand, I ate noodles and vegetables for breakfast. When I was in Japan, I ate rice, nori (seaweed), and salted, preserved fish and soy sauce for breakfast. When I was fourteen years old and went to summer school in Germany, I ate ham, salami, and cheese with bread for breakfast.

Why do we have such a narrow idea about what to eat for breakfast? Why, in America, are omelets usually only served for breakfast? I have eaten them with salad for lunch in France and London. And then, some families drink coffee with their lunch or dinner, and we Brits like to eat Heinz baked beans for breakfast on toast, or in a baked potato for lunch! I will give you some information next, but I want you to start by remembering that you don't have to eat what you grew up thinking was "good" food for each meal based on your culture. *Capisce?*

WHEN TO EAT AND WHAT?

What I am going to tell you here, in the simplest terms, is when to eat what and what the what is! First, we are back to the Ayurvedic Clock (Figure 4).

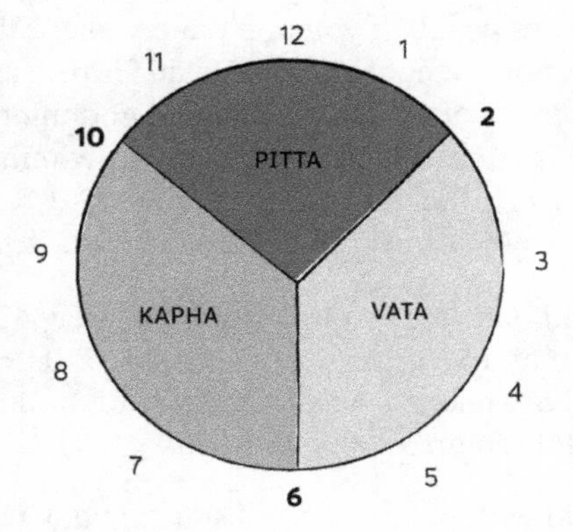

Figure 4. The Ayurvedic Clock.

> Don't eat if you're sad or angry. Your body won't digest it well!

As I mentioned in Chapter 2, your biggest meal should be between 10:00 a.m. and 2:00 p.m. for a number of reasons. The first is, this Pitta timeframe is when you have the most digestive fire (agni) or in western-speak, bile in your system. (Again—don't you just love this fantastic melding of eastern and western science?!)

> Eat your biggest meal of the day between 10 a.m.–2 p.m. when your digestive fire is strongest!

Secondly, your system will most probably be upright at this point and will continue to be upright for a few more hours. So, you have gravity on your side, which helps move food from the stomach to the small intestine to the large intestine.

Thirdly, your system is awake—that is, hopefully you've pooped, exercised, and eaten a nutrient-dense but easy-to-digest breakfast so that your alimentary canal is in peristalsis (out of detox mode—remember that every night, if you eat early, your bod will detox) and totally ready for an influx of tastiness.

Look at the clock again. We tend to eat our other meals in the 6:00–10:00 a.m. (Kapha) and 2:00–6:00 p.m. (Vata) slots, but those are not beneficial for eating big, hard-to-digest meals. Remember,

Kapha is associated with water and earth (dense, slow, and heavy) whereas Vata is associated with air and space and cold. Pitta is fiery, which is why we want to eat our biggest meal at lunchtime, and we want to be sleeping by 10:00 p.m., so that fiery energy is used toward detoxing our minds and bodies—not keeping us awake to party, argue with people on social media, fold laundry, or pay bills. Vata time and Kapha time are not great for digesting heavy meals. Hence, eating light, easy-to-digest meals for breakfast and dinner is the smarter way to go.

Having trouble falling asleep? Eat a sweet potato at supper time.

What to eat? Plants. Base your meals around plants. Eat plants first and foremost. Add spices and herbs to make this process more interesting. How do you make the shift?

1. Start embracing spices and herbs. Squeezing fresh lemon juice and chopped cilantro or parsley over your steamed broccoli will make it taste more interesting. Sautéing your zoodles (zucchini noodles, or as they call it in Britain, courgetti or courgette-spaghetti) in coconut oil or veggie stock with a little garlic and cayenne will add a boatload of flavor with not much effort. Make some dressings (I love the ones that Jane Esselstyn came up with or repurposed for Rip Esselstyn's books—e.g., combine two tablespoons of hummus, two tablespoons of balsamic vinegar, two tablespoons of orange juice, two teaspoons of mustard, and one teaspoon of grated ginger. I make it in bulk and keep it in a jam jar in the fridge).

2. Commit to eating just plants for one meal a few times per week, e.g., every other day. No animal products! Just try!

3. Explore new plants. Walk around your grocery store or farmers' market each week and buy at least one new vegetable or fruit (or get your kids to pick them out), and then learn where it comes from and how to cook it.

4. Bulk cook. As my friend and fellow coach Harriet says, "Cook once, eat at least twice." Make it easy on yourself!

TO SNACK OR NOT?

Your digestive tract is very long. But its focus and attention tends to be like my dog's, meaning it goes toward the most interesting—

squirrel! Er, what were you saying?—new thing you present it with. Say you eat oatmeal with banana and seeds at 8:00 a.m. Your food is hopefully in the small intestine by 10:30 a.m., when you feel peckish and eat a piece of toast and peanut butter. The blood, focus, attention goes to the stomach, and the toast and peanut butter—this new food—is way more interesting than the movement of your oatmeal along your small intestine (and eventually into your colon). This slows down the digestion of your oatmeal, which means that you are more likely to get constipated. Who wants that?! Not *moi*, darling, if I can possibly avoid it!

> Avoid eating fruit after a big meal—it tends to ferment and cause distension and discomfort.

Now, if you're really, really hungry because you have been exercising either your mind or body (thinking and using your mind also makes you really hungry, because your brain requires energy, too), then you can try a couple of things.

First, drink a large glass of water, herbal tea, or make your own digestion-stimulating brew of hot water with coriander, cumin and fennel seeds and a knob of lightly crushed fresh ginger. Wait. Just wait. Your body often confuses hunger with thirst.

> Making kombucha is easy! Get a scoby (symbiotic culture of bacteria and yeast) and start! Try apple sauce and cinnamon sticks for a kid-friendly and deliciously fizzy brew!

Second, if you're still hungry, really hungry, then try a piece of fruit. Fruit can ferment when eaten with other foods or after a big meal, so just notice that and snack on fruit only.

Third, reflect on your prior meals. I invite you to notice your food choices and what happens when you eat different foods. If you eat a higher-fat food for breakfast, like avocado, does that stave off the hunger for longer? If you add more protein at breakfast by eating a tasty big bowl of oatmeal with chia seeds (12.8 g of protein, as noted above), does that work? That staves off my hunger for the morning! I also have way more energy if I up my intake of greens. The easiest way to do that is to juice kale, spinach, etc., with some carrots and apples for breakfast! Did you eat well for lunch? Or did you just eat a few salad leaves at noon, so it's no wonder you're starving by 3:30 p.m.?

Generally, you want to eat sufficiently well to avoid snacking. Remember that feeling a bit hungry means you'll REALLY enjoy your next meal. It won't kill you to feel a little hungry. Learn to sit with it. Not like a martyr, but just, like, "Huh, when I am hungry, I feel (fill in the blank)."

Caveat: This does not apply if you're a young sprog (kid) who might get very hangry (hungry plus angry) or have blood sugar issues or other conditions that require more frequent eating.

SUMMARY

I've given you a lot of information and things to ponder in this chapter. Some of it was facts—I always think it's good to know the facts, because knowledge is power. Also, some of it was about mindset—what you were taught as a kid, what you fear, and what it means to nourish oneself (which needs some pondering, in my view). Did you notice that the definition of nourish has nothing to do with consuming food?

WHAT DO YOU NEED TO UNLEASH YOUR POWER?

Here are some actions you can take to make food and eating more about nourishment and pleasure than scarcity and ill health. Also, here are a few things you can do to help make these changes stick—both practical tips and mindset shifts and support.

- **Storage for your meal prepping**: Invest in some good glass Tupperware so you can prepare and store your veggies and fruits. Keep your glass jam jars for making dressings and storing soups.
- **Inspiration:** If eating more plants is a new thing for you, then start to explore some new recipes. Use the recipes at this back of this book or join a vegan meetup group, which explores local vegan and/or vegetarian-friendly restaurants or has cooking classes. Want more inspiration? I love these books:
  - *The Oh She Glows Cookbook: Over 100 Vegan Recipes to Glow from the Inside Out* by Angela Liddon
  - *Thug Kitchen: The Official Cookbook: Eat Like You Give a F*ck*
  - *Eat-Taste-Heal* by Thomas Yarema
  - Any of Rip Esselstyn's books (e.g., *Plant-Strong: Discover the World's Healthiest Die*t, which was originally published as *My Beef with Meat*)

    - *The Plantpower Way*: Rich Roll and Julie Piatt

- **Be gentle with yourself:** For most of us, this means making the change slowly and sustainably. If you currently drink two to three sodas a day, then cut down to one a day, right now. Then next week, drink one a week, then during week three—nada. Replace soda with naturally-flavored water or soda water ("natural" here means with a wedge of lemon, cucumber, orange, or mint leaf, etc.)! If you eat animal products at every meal, start with making breakfast animal product-free. Try making supper plant-based (your sleep will be way better)—but easy does it!
- **Cultivate support from your people**. Talk to your partner or spouse, coworkers, kids, and parents about your commitment to shifting to a healthier, plant-based diet. Don't make it a big deal. See if they will explore with you, like the "Meatless Monday" movement, or at least ask them not to denigrate your choices. Perhaps create a food-share with some local friends, where you cook for three families once a week, in return for them cooking for you (that will give you three meals covered each week). Bulk cooking veggie stews, bean chilis, trays of roasted root veggies.

THE MAIN MESSAGE

- Eat more plants, with your biggest meal being lunch.
- Look at your plate and fill three-fourths of it with plant-based foods. If that seems really hard, shift it week to week...one-fourth plants, to half plants, to three-fourths plants!

How?

Plan your meals. Set aside time to bulk cook. Remember this saying: "Cook once, eat at least twice."

1. Put your meal times in your smartphone calendar, on repeat. Write Post-it notes and paste them around the house and/or office. Reschedule and make space for meals—even if it's just fifteen minutes of you quietly eating alone!

2. Track your progress and be your own habit scientist. Use a mobile app and/or write down what you eat and when. Take note of how you feel when you do or don't follow these suggestions!

"Let medicine be thy food and let food be thy medicine."

—Hippocrates

TIME OUT

Are you feeling that resistance again? I know, right? Food is a big freaking deal. I totally hear you. There are so many emotions tied to it and we use and abuse it—rewarding ourselves for our achievements and eating or starving ourselves when life gets stressful. Food is vital for our life—it's our fuel. It's sensual and delicious and fun to smell, taste, touch, see, and hear what you're cooking, growing and sharing with friends. The key is to face it, organize it, and then ENJOY it! Because you know you want to unleash your power and become a freaking rock star. Here are two tips that will help you.

Tip 1: Rewards

One of the things to really pay attention to is that you don't use food as a reward. That's a slippery slope of no-bueno-ness that you totally want to avoid. A fellow coach friend of mine, Elijah, frames it this way. She says, "Stop using the word 'treat' for those sugary, salty, fatty things like cupcakes or a bag of chips. They are not a treat, they are an unhealthy indulgence." Most of us have felt the benefit of the reward of the habits that we are putting in place—e.g., eating smaller, lighter suppers without a lot of booze creates a deeper and more nourishing sleep. But, it can take a while for these to start working and for us to really notice that it's the new habit that's causing this feeling of awesomeness.

So, we sometimes need to give ourselves a reward that has nothing to do with the activity, similar to the way you might give your toddler a sticker for making it to the potty on time! An example? "If I eat dinner by 6:00 p.m. four nights this week, then I am going to book myself a massage for next weekend." Food and massage—not an obvious pairing, but you're connecting something you want (the massage) with something that you know will make you feel better (eating a smaller, lighter supper). Give it a shot. Like a toddler, I like to tie my rewards pretty intimately with the event and not wait for three or four or five events. That's how I help myself develop great habits!! But then, this is personal, so know thyself.

Tip 2: Kintsugi

Bulk cook on Sunday afternoon and a weekday or evening which is not too busy.

✦ Chop up veggies—peppers, celery, carrots for hummus and dips—and dice up some tomatoes, onions, peppers, and cilantro, and box up.

✦ Make naked layer salads in Mason jars and a few separate jars of vinaigrettes.

✦ Make a batch of chili, beans, vegetable soup.

✦ Roast a tray of root veggies and box up in fridge.

✦ Cook a rice-cooker's worth of grains/lentils/beans with a bouillon cube (quinoa, black beans, couscous, brown rice, barley, amaranth).

✦ Then you can create easy bowls of food during the week, e.g., spinach, rice, roasted veg, chopped peppers, and cilantro with a vinaigrette.

Sometimes things will go wrong—you'll get sick and eat processed crackers for a week and then pizzas and and and.... Maybe a relationship will end, you'll lose a job and eat a tub of ice cream and drink a bottle of wine. Forgive yourself, let go, and then remember that these events do not define you. Have you ever heard of Kintsugi? It's a Japanese art in which broken pottery is glued back together by using gold glue, so you can see and celebrate the breaks.

So, remember that if you feel broken, if you feel like it's all over, have a wail and a scream and cry, but then, for goodness' sake, get off the floor, turn on your favorite song, have a shimmy and shake it all over the place, and start again. Start anew.

Bonus: When you're on vacation, at a party, away for the weekend, and your eating is different—notice how you sleep, feel, and digest.

WHY? If you eat better, you'll sleep better, work better, have more energy, and love your life.

HASHTAGS #SAD #StandardAmericanDiet #GoodFat #NotAllCaloriesAreEqual #DietsSuck #Endothelium #Plantstrong #PlantstrongPantstrong #AvoidProcessedFood #EatSmallerPortions #EatMorePlants #EatPlants #SerotoninInYourGut #EatLunch #EatABigLunch #MainMealIsLunch #PlantBasedDiet #PittaTimeIsLunchtime #UseSpices #BulkCook #DigestiveFire #Agni #Bile #CookWithHerbs #EatYourNeighborhood #AlimentaryCanal #AvoidSnacking #Avocado #DigestionTakesTime #SnackOnWater

CHAPTER 6

EXERCISE

"True enjoyment comes from activity of the mind and exercise of the body; the two are ever united."

—Wilhelm von Humboldt

This is where I tell you to get your butt off the coffee shop stool and into the gym…yeah, yeah, yeah. I know you know this. But seriously, did you know that the longer you spend sitting on your butt, the bigger it gets?! Really, it does! The tissues in your buttocks actually break down, which causes spreading and thus your butt gets bigger. I like to unleash my power with a nicely exercised butt, you? GET UP!

"Those who think they have not time for bodily exercise will sooner or later have to find time for illness."

—Edward Stanley

WHY EXERCISE?

I often go to the CDC for statistics for my clients and my talks because it provides large-scale, government-collated data about the health of the nation. These statistics can provide a helpful baseline. So, when I was collecting data about the state of the nation for exercise, I found that you can even download PDF's from the CDC containing guidelines for exercising—one for walking in malls, and the other for walking in airports! See? You can exercise anywhere!

When I first moved to America, I lived in St. Louis, Missouri, and I would see women power-walking around the mall, with weights. I figured it was because the summers often had temperatures of 105 degrees Fahrenheit and humidity of 60 percent, giving it a feeling of about 150 degrees Fahrenheit…but perhaps it was also because they had been given a mall-walking exercise plan!

AGAIN, WHY EXERCISE?

Physical activity will make you feel better, because your eleven body systems (circulatory, excretory, digestive, nervous, immune, etc.) will work better individually and as a team. People who shake it on a regular basis tend to live longer.

Inactive adults have a higher risk for early death, heart disease, stroke, type 2 diabetes, depression, and some cancers. Nice, isn't it? In addition, getting regular activity can help with weight control, and seems to improve academic achievement in students.

So, here's where we are in the Divided States of America. According to the CDC, about one in five adults (21 percent) aged eighteen to sixty-four years met the 2008 Physical Activity Guidelines, which are pretty easy to master: two-and-a-half hours per week of moderate activity or one-and-a-quarter hours per week of vigorous activity and two days per week of weights. Less than three in ten high school students get at least sixty minutes of physical activity every day.[46]

Interestingly, but not surprisingly, you're more likely to meet these guidelines if you're young, male, non-Hispanic, do not live in the south, live above the poverty level, and have more education.

Exercise will also make you feel better and be more productive! When you exercise, your body produces endorphins, which are the "feel-good" hormones. They have the same effect on the brain as does morphine. Also, regular exercise promotes, as I noted above, lower rates of depression. This is huge, even for people who are not clinically depressed. Because endorphins act like analgesics, one's perception of pain is also reduced by exercising, which is massive.

I find my mind works better, too, when I am exercising regularly. I suspect it's because I am moving the blood around my body and brain and I am sweating, which flushes out the toxins! Also—I can feel my general levels of energy are SO MUCH HIGHER when I exercise regularly. I want to show up and be the best version of myself, unleashing my power left, right, and center towards my kids and family, my business, my friends, my hobbies, my passions, and charitable concerns!

Also, if you're feeling lazy, you might like this stat: For every pound of muscle you have, you burn fifty calories a day at rest.

---

46 "Facts about Physical Activity." Centers for Disease Control and Prevention. May 23, 2014. http://www.cdc.gov/physicalactivity/data/facts.htm.

That's inspiration for me to lift weights regularly! That my body burns all those calories when I am doing nothing is a TOTAL bonus! When I wrote the first draft of this book in 2017, I could leg-press 418 pounds. Just sayin'!

"Physical fitness is not only one of the most important keys to a healthy body, it is the basis of dynamic and creative intellectual activity."

—John F. Kennedy

EXERCISE DOESN'T HAVE TO BE A MASSIVE HOLLYWOOD PRODUCTION

Keep an exercise diary. Write down what you do and how you feel. Notice the patterns (e.g., going to lift weights made me feel strong, but I only like it when the trainer tells me what to do, not when I am doing it on my own, or when I am in a class listening to loud music with lots of people around me).

If you're like me, and you get bored easily, then mix it up, but with clear consistency about what works for your body and mind. I am hypermobile, so I need to do weights so that my muscles hold me together. Also, if those CDC stats about two-and-a-half hours per week make you think, "Hell no, I can't find that time," then keep reading, my friend!

Yoga is what keeps me sane, strong and flexible in my body and mind. After practicing Vinyasa yoga for years, I have now added Iyengar yoga to the mix because it teaches me to hold poses for longer, which is a challenge for me (and you can probably tell that I like to focus on what I find hard, so I can evolve!), and I love my teacher and the students!

But I also get bored and love music, so I have a rebounder (mini trampoline) in my sitting room (irony here, anyone?) and I crank up the tunes and bounce a few times a week. Note: a jump rope will also do! Just make sure your ceilings are high enough and your chandeliers are out the way, darling! I also love to listen to podcasts and sassy-girl tunes and so I go running about once a week, with my rescue pup, which keeps us both fit!

I've recently joined a gym so I can swim, following an injury. I've also taken ballet and barre classes and I've been meaning to try pole dancing!

What I am saying is, you do you.

If you need accountability, set it up.

If you need variety, set it up.

If you need music, set it up.

If you need to do it at home, research mobile apps and websites.

If you need group energy, research gyms and studios.

If you need predictability, do Bikram or Ashtanga yoga, run a familiar track.

If you need variety, do Iyengar or Vinyasa yoga.

Maybe just getting rid of your car and taking public transport will be enough—particularly if you track your steps using your phone, a smart watch, or an activity tracker.

Got it?

> Do a free/or low-cost week at your local yoga studio or gym! Get a flavor for what you enjoy.

Just don't put this on the back burner, because getting the prana—the energy—flowing will make the difference between a successful you and an unsuccessful you! Track it; make your commitment visible and do it every damn day.

There are a lot of different ways to move your body, so make sure you try out lots of things, especially if you haven't found activities you enjoy yet. If you're a member of a gym, try out all the classes, like spinning, weight classes, Zumba, or yoga. Also, sign up for a personal training session (most gyms offer a free trial), take notes of what they get you to do, and ask lots of questions!

Below you'll find some more suggestions of things that you can easily do at home (or even while you're staying at your in-laws' house or at a hotel for a conference).

INDOOR (FREE OR MINIMAL COST)

- *The Scientific 7-Minute Workout*. This has twelve exercises that each take thirty seconds. You can download the app or Google the *New York Times* article about the 7-Minute Workout to find the two sets of exercises they now have. You can repeat it a number of times, if you have the time. NEED: Floor space, a clear wall, and a sturdy chair for stepping onto.
- *High-intensity interval training (HIIT)*. This is a variety of short (ten- to twenty-minute), high-intensity movements such

as lunges and burpees. YouTube has a number of channels. NEED: Floor space, and some of the activities require weights (I use cans of black beans as a replacement!).

- *Rebounder (mini trampoline) or skipping rope.* Do this for at least ten minutes (remember, some exercise is better than none). NEED: Rebounder or skipping rope, floor space, high ceilings.
- *Yoga.* If you don't know what you're doing, check out a YouTube channel or join YogaGlo (a subscription service for yoga classes—not advisable for a yoga virgin). I suggest going to classes, too, even if you're not a yoga virgin (I have been teaching yoga for about ten years), so you can have a certified teacher help you with correct alignment to prevent injury. Now, don't get hung up on it needing to be ninety minutes to be worthwhile. Even just five minutes of sun salutes will make you feel more like a human being! NEED: yoga mat, floor space.
- *Run up and down your staircase for ten to twenty minutes.* Put on some funky-ass tunes or a podcast and get to work. This one's great for your butt and thighs. Who doesn't want a nicely sculpted derrière?! NEED: staircase.

OUTDOOR (FREE OR MINIMAL COST)

- *Run.* Lace up your sneakers, choose some good tunes or a podcast, and off you go. If you're new to running, start with a small goal such as running for the length of your favorite song. Then build it up, aiming for twenty minutes. I like the Track My Run app because it talks to me and tells me after each mile what my speed was (I'm not that impressive—I limit myself to two miles). It's fun to be a nosy neighbor—check out whose gardens are nicely tended, who has some quirky trim color, whose roof needs replacing, or who has a crush-worthy front porch (I've always wanted a porch and porch swing). Alternatively, find a public track (e.g., at a local high school), or a nature reserve, if you like it more outdoorsy. NEED: Good sneakers (replace them every 500 miles or so), good sports bra to hold your boobs in place or tape over your nipples if you're a guy—to prevent chafing—and comfortable pants and top (nothing will put you off more than chafing of the inner thighs or armpits due to ill-fitting clothes!).

- *Walk*. Same as above. Just slower. More time to be nosy or listen to the birds tweeting, or the wind blowing in the trees. Whatever is your thing. Make it happen. Also, start to add more walking to your daily life—park a ten-minute walk from your destination, take public transport to work or events, and take the stairs rather than the elevator. NEED: Good supportive shoes. A dog will help—you have to walk your pup every day, so that helps make this a habit.
- *Bike*. Okay, this activity can cost a lot if you buy a swanky bike and load up on all that Lycra. Or you can be like me and get a re-conditioned bike, pull on a pair of shorts or leggings, and use it as a mode of transport. Road bikes are lighter and fast moving (I have a 1980s red road bike I picked up from a second-hand bike store), or get on a bigger mountain bike and explore the woods. NEED: Bike and helmet. Protect your noggin with a good helmet (replace if you've had an accident). It's full of your brains, which are a good thing to protect.
- *Cross-country skiing*. In my part of the world (northeast Ohio), you can rent some cross-country skis pretty cheaply and head off in one of our Metro Parks. It looks easy, but it's a really freaking intense workout in beautiful surroundings! First time I did it, I thought my jacket had lost its waterproofness. But no, I had sweated through my roll-neck, fleece and jacket! NEED: Skis and poles (rent or buy), a track.

Try out a personal trainer for a session and take notes about what you should do for your specific goals!

This list could go on and on—I could tell you to try paddle boarding (which is fricking amazing for your core and is lovely and peaceful), rowing, kayaking, skiing, swimming, rock-climbing, sailing, etc. The point is, just find a way to move every day. Find people to do it with. Make it a routine. And enjoy yourself, for goodness' sake!

SUMMARY

So, hopefully, you're convinced that exercise is a good thing to do. You'll feel better (in your mind and body), you'll be helping to protect yourself from long-term health issues, and you know, you might even look better and find that your jeans slide over your butt a little better if it's toned. Just sayin'.

The biggest part of this is really starting to learn what works for you—your day, your life, your body, your mind—and then doing it. Every. Damn. Day. Got it?

WHAT DO YOU NEED TO UNLEASH YOUR POWER?

> Want to lose weight? Work out in the morning. Want to perform at your peak in your exercise practices? Work out in the afternoon (Daniel Pink).

Time, stuff, and commitment. It's that simple.

TIME: Pull out your calendar and schedule exercise times. The earlier in the day, the better your day will go and the less likely you will be to avoid it. Daniel Pink notes that exercising early in the day is more beneficial for weight loss, and exercising later in the day is better for productivity. I like to mix it up, but be consistent with the days. For example: do yoga on Thursdays, swim on Wednesdays, weights on Monday, HIIT videos Tuesday, and go for a run on Fridays.

STUFF: Make sure you have what you need to do it—a yoga mat, a sports bra, sneakers, etc. Learn to pack it up the night before—put your yoga mat by the back door or in the car and your sneakers and gym clothes on the floor by your bed, for example.

COMMITMENT: This is both the simplest and the hardest. Put your exercise times in your calendar and behave as if you're meeting your hero, i.e., DO NOT FAIL TO SHOW UP. Join people for some of it—yoga, gym classes, runs, or walks—because accountability is super helpful.

THE MAIN MESSAGE

Exercise every day, even if it's just for five minutes on your trampoline or a few hops with a jump rope. Know yourself: Do you like groups, music, or repetition?

How?

Try the following to put these ideas into action:

1. Reverse engineer/plan your weekly exercise, remembering that it's easier to establish and maintain the habit if it's roughly the same every week.

2. Put your exercise commitments in your smartphone calendar, on repeat. Write Post-it notes and post around the house and/or office. Tell your housemates and/or friends what you're doing—set up your accountability partners.

3. Track your progress and be your own habit scientist. Use a mobile app and share on social media, so you create momentum and engagement. Write down what you do—and celebrate the wins (e.g., three yoga classes this week). I like to hang a calendar in my bathroom (where I obviously visit regularly, because I eat a lot of plants and I like to check whether I've got broccoli between my teeth!) and track my goals. Take note of how you feel when you do or don't follow these suggestions. During the writing and editing of this book, my leg press weight from 385 to 418 pounds! Ahh, the satisfaction brought about by tracking!

TIME OUT

STOP in the name of (self) love, before you break your heart. I know exercising can feel like a duty sometimes, and I am all about making your heart unbroken, which means eating well, sleeping well, and exercising well. And yes, of course, there will be those days when the alarm beeps at 5:30 a.m. and you feel like smacking it and rolling over. This is such a vital habit to develop properly and commit to if you really want to unleash your power!

But you know you'll feel so much better in that meeting with that irritating person when your body has had a good dose of endorphins and your belly doesn't hang over your waistband. Just sayin'. Sometimes on those days, plan a reward for yourself that has nothing to do with the exercising. For example, "If I exercise today, I will buy one of those swanky *matcha* tea lattes on the way to work." Sometimes we need a little something-something, you know?

Tip 1: Sometimes It's About Faking It Until You Make It

Wait, wait, wait—hear me out. Did you know that if you grip a pencil between your lips, it makes you happier? That is, your body can change your mind, your mind can change your behavior, and your behavior can change your outcome. It's not just that your mind changes your body. This is a two-way street, baby, and you've gotta learn how to drive both ways!

What would happen if you just tried exercising every day this week? Just try! Notice how you feel when you exercise more consistently. FYI, it's easier to do something every day to make a new habit stick, so I suggest committing to exercise EVERY day. Make it small, doable, and fun!

Tip 2: Accountability

> Get an accountability partner—a girlfriend who rings your doorbell at dawn to go for a run with you! A friend who meets you on the tennis court (and won't be able to play without you)!

Having a partner, who eggs you on (to do the GOOD habit) can be super helpful, to keep you on track and help you celebrate the wins, as well as help you keep at it when you feel like it's simply the most boring and uninspiring thing to be doing.

I like to be accountable to myself, —both digital calendars which beep and ding and remind me, and the handwritten visual aids in front of me, where I log my achievements. Post-it notes to remind me to go to gym, to meditate, whatever it is, and then a big, colored chart where I can check off what I am doing and then I can give myself a "hell yeah" when I hit my goals—it's SOOO satisfying to see it all written up there in bright colors! Also, your brain processes things more efficiently when it's handwritten!

BONUS: Whenever you ski, hike, bike, swim, rock-climb, etc. for many more hours during the day than normal (e.g., if you go on a sporty vacation, or on the weekends), notice how you sleep, what kind of food you crave and find yourself eating, and how you feel in general.

WHY? If you exercise, then you'll sleep better, eat better, work better, have more energy, and love your life.

HASHTAGS #ExerciseDaily #YogaEveryDamnDay #Endorphins #Yoga #Running #Weights #Walking #TrackYourSuccess #FakeItUntilYouMakeIt #Accountability

CHAPTER 7

RELATIONSHIPS WITH YOURSELF AND OTHERS

This book has focused on how to establish the key habits to create a life of freedom. But a massive element of this is relationships. So, now, in this juicy section, I am going to get you to start thinking about relationships.

This is about both the relationship you have with yourself (the categorization of yourself from others, in terms of what your boundaries are) and secondly about how you engage with others (what your relationships look like, how you treat others, who you have sex with, who you live with, how you interact with those in your life, and how you manage this). So, your relationships with yourself and with others are intertwined.

A lot of books and theories about relationships with others and with yourself are out there, many of which I have read and pondered. Some of the ones I have enjoyed include:

- *Grit: The Power and Passion of Perseverance* (Angela Duckworth). I met her when she gave a talk at a school in London and she is spectacular in the flesh—witty, smart, and self-deprecating. #GirlCrush
- *Turning Your Mind into an Ally* (Sakyong Mipham). My copy of this is so well-worn, because it's been carried around, read, and underlined (sacrilege in the home I was raised in, but for this bibliophile, it's a sign of true love).
- *The Wisdom of No Escape and the Path of Loving-Kindness*, *Comfortable with Uncertainty*, and *When Things Fall Apart: Heart Advice for Difficult Times* (all by Pema Chödrön). Simple, clear, undeniable brilliance about your relationship with your mind.
- *The Power of Habit: Why We Do What We Do in Life and Business* and *Smarter, Faster, Better: The Secrets of Being Productive in*

Life and Business (Charles Duhigg). Fun and page-turning (he's a journalist) with lots of useful practical tips, all grounded in stories of other people.

- *Nudge: Improving Decisions About Health, Wealth and Happiness* (Richard Thaler and Cass R. Sunstein). This book made me exclaim out loud, a lot (I tend to do that when my mind or physical senses are receiving stimulation; you'll know if I like your cooking!)
- *Big Magic: Creative Living Beyond Fear* (Elizabeth Gilbert). This is about the creative process and how you have to be diligent and create structure if you want the creative energy to flow. Love her vulnerability. In person, she is witty, vital and takes no prisoners. I went to a workshop that she and Cheryl Strayed did on creativity and writing and it was fantastic!
- Literally anything by Brené Brown. (Side note: the audiobook of *Rising Strong: How the Ability to Reset Transforms the Way We Live, Love, Parent, and Lead* made me weep. Hearing her read her own words in her lovely Texas accent was so fab.)
- *Nonviolent Communication: A Language of Life* (Marshall B. Rosenberg). The kind of conflict resolution (think Hutu: Tutsi; Israel: Palestine) presented in this book challenges you deeply to face yourself and how you judge others. It discusses how it'll be really hard to have meaningful conversations with people who have very different views from you if you can't master deep, active listening.
- *The 5 Love Languages: The Secret to Love that Lasts* (Gary Chapman) presents the idea that how you communicate love to the important people in your life may not be how you like to receive it (e.g., I like to be hugged hello, kissed on the cheek, and have arm around my shoulder—the Physical Touch love language—and I also like to be told "Thank you," and "I appreciate you"—Words of Affirmation-—and I really appreciate it when my loved ones do Acts of Service, such as emptying the dishwasher).
- *Conscious Loving: The Journey to Co-Commitment* (Gay and Kathlyn Hendricks). This book challenges you to consider how to commit to someone else without losing yourself.
- *Vagina: A New Biography* (Naomi Wolf). This deeply personal, political, medical book touches on creativity and orgasm, and

(like the book below) pushes us women to understand and embrace ourselves from the deepest levels.

- *Pussy: A Reclamation* (Regena Thomashauer). This book presents the notion that a woman's power is grounded in the deepest sexual core and that we need to embrace, understand, and act from here to be successful. When women take charge in stereotypically "masculine" places, such as in business, it can be easy to stop acting from a place of feminine energy. This book debunks that myth.
- *The Life-Changing Magic of Not Giving a F*ck : How to Stop Spending Time You Don't Have with People You Don't Like Doing Things You Don't Want to Do (A No F*cks Given Guide)* (Sarah Knight). This was a divorce/new life/new house present from a girlfriend. Totally what I needed because it lays out a step-by-step guide!

This topic is huge and fantastic and of vital importance when you start on this path of self-evolution. However, I am not going to delve deeply into it here—that's a topic for another book! What I will do is present some ideas to start considering around this topic, so that you can start to ponder and have conversations with the important people in your life.

But first, no relationships—with yourself or with others—will work without meditation, IMHO.

MEDITATION

> Meditate. It's the single biggest game-changer for mindset! Use an app for accountability and guidance, e.g., Insight Timer, 10% Happier, Headspace, or Deepak Chopra (love his voice—I find it very soothing).

Here is where I am going to get super bossy, like holding that cat o' nine tails in my right hand and slapping my left palm with it, so you know I MEAN BUSINESS. Got it?

Meditation is necessary to have good and nourishing relationships with yourself and with other people. Why? Meditation is the single most important thing you can do to shift your relationship with yourself and with others. That's because it requires you to sit with your mind and become friends with it. Terrifying thought, right?!

"Meditation is about seeing clearly the body that we have, the mind that we have, the domestic situation that we have, the job that we have, and the people who are in our lives. It's about seeing how we react to all these things. It's seeing our emotions and thoughts just as they are right now, in this very moment, in this very room, on this very seat. It's about not trying to make them go away, not trying to become better than we are, but just seeing clearly with precision and gentleness."

—Pema Chödrön, *The Wisdom of No Escape and the Path of Loving-Kindness*[47]

Meditation teaches you about your own mind. You can't have a good relationship with yourself if you don't know what you're dealing with, and you can't have good relationships with others if you don't know yourself.

Meditation also teaches you to let go. By practicing letting go of the thoughts that irritate, frighten, and upset you about yourself and others, and by coming back to the breath, you will realize that the thoughts and emotions that travel across your mind do not define you—they are simply like the ever-changing weather. If you do not practice letting go of these things, then you won't have good relationships.

Meditation teaches you to pause before you react. This is great if it's for your behavior—pausing before you eat that third wedge of pizza or sharing a pissy social media post. It's hugely important when you're pausing before you shout at your kid, react with irritation to your partner, or say yes to another ridiculous and unfair demand from your boss, mother, or *quelqu'un d'autre* (someone else)—sorry, I slip into Franglais sometimes—it happens when you grow up in Britain.

Meditation can change your brain. One of my favorite studies was performed by Richard Davidson at the University of Wisconsin-Madison, in which sixteen Tibetan Buddhist Monks (all of whom had practiced at least 10,000 hours of meditation—a working definition of expert) were compared to sixteen age-matched controls. The study involved looking at the activity in their brains by using functional magnetic resonance imaging (fMRI) in response to neutral, positive, and negative emotion-inducing sounds while the novice and experienced meditators were practicing a kind of meditation called compassion, loving-kindness or metta meditation.[48]

---

47 Chödrön, Pema. *The Wisdom of No Escape and the Path of Loving-Kindness*. Boston, MA: Shambhala, 2010.
48 Lutz, Antoine, et al. "Regulation of the Neural Circuitry of Emotion by Compassion Meditation: Effects of Meditative Expertise." *PLoS ONE* 3 (2008). http://journals.plos.org/plosone/article?id=10.1371/journal.pone.0001897.

In this practice you wish people physical and mental well-being to specific people—someone you love, dislike, feel neutrally about, and yourself. I found both this practice and another, called Tonglen, in which you absorb the pain of others, increased my compassion for others, and was really helpful to me when I had moments of such hurt and anger with my ex-husband as we navigated the legal and lawyer-involved divorce process.

Needless to say, the data showed that the development of the mental skill to cultivate positive emotion changed the activation of the brain circuits which are associated with empathy and the response to emotional events.

Meditation is a hugely beneficial practice because it improves your ability to be kind to yourself. In a 2012 study, 77 long-term meditators were compared with 79 non-meditators in terms of their psychological well-being, mindfulness, and self-compassion.[49] Psychological well-being was significantly correlated with long-term meditation. We all know how we tend to be much nastier to ourselves than to others. When we are more compassionate, the way we interact with others changes for the good. We boost our ability to empathize, which makes it easier for us to extend help where needed and make decisions that will benefit all, not just ourselves.

Practice equal breath, making the inhale and exhale equal, bringing yourself into harmony.

Meditation makes you happier—it's been shown to reduce depression, anxiety, and symptoms of PTSD. For example, a study of military members that compared 37 meditators to 37 non-meditators showed massive improvements in those who meditated: After one month, 83.7 percent of the meditators had stabilized, decreased, or ceased taking psychotropic medications for their PTSD and 10.3 percent had increased their medications.[50] In comparison, the non-meditator group had 59.4 percent stabilized, decreased, or ceased medications, and 40.5 percent of participants had increased their medications. Six months later, there was over 20 percent difference in psychological symptoms between the two groups. These data were statistically significant. Pretty rad stuff, isn't it? Meditation is the BOMB!

49 Baer, Ruth A., et al. "Mindfulness and Self-Compassion as Predictors of Psychological Wellbeing in Long-term Meditators and Matched Nonmeditators." *Journal of Positive Psychology* 7 (2012): 230–238.
50 Barnes, Vernon A., et al. "Impact of Transcendental Meditation on Psychotropic Medication Use Among Active Duty Military Service Members With Anxiety and PTSD." *Military Medicine* 181 (2016): 56–63. http://militarymedicine.amsus.org/doi/full/10.7205/MILMED-D-14-00333.

So, if you're happier, less depressed, more compassionate, less anxious, and so on, this makes your life more pleasant, right? And if you're happier, your relationships will be better. Did you know that happiness is contagious? No, really, it is—and there's research to support this. If you're happy, then it will spread up to three degrees of separation (i.e., other people). The effect decreases with time and distance.[51] When you're happy and you treat those around you well, that happiness spreads!

Meditation improves your immune function by changing your DNA. At the ends of your chromosomes are telomeres, which shorten as you age, get sick, and so on. In studies that analyze people who meditate, there is increased telomerase activity (the protein that adds to the telomeres). For example, in a 2011 study, the data suggest that meditation increased people's sense of perceived control and purpose in life and decreased their negative feelings, which in turn boosted their telomerase activity, with implications for telomere length and immune cell longevity.[52]

The telomeres at the ends of your chromosomes get longer when you meditate, which makes you healthier. This is good for you and the work you do in the world—who doesn't want to be healthier?! And, like the concept of vaccination, if you're getting sick less often, you are not spreading lurgies, a.k.a. germs (something I am super aware of as a mother of three school-age kids who often come home with colds, flus, and stomach bugs)!

Meditation creates more balance and healthy modifications in your body. For example, it can improve lung function and reduce blood pressure (implicated in heart disease, kidney disease, and dementia) and cortisol levels (which we know is associated with stress, reduced immune function, and belly fat). A study compared meditators to non-meditators and found that after meditating, serum cortisol levels, blood pressure, and pulse rate were significantly reduced, and lung function was significantly improved.[53]

If your overall functioning is good, that benefits you and your relationships with others. It's super stressful living with a chronic

---

51 Fowler, James H., and Nicholas A. Christakis. "Dynamic Spread of Happiness in a Large Social Network: Longitudinal Analysis Over 20 Years in the Framingham Heart Study." BMJ 337 (2008). http://www.bmj.com/content/337/bmj.a2338.

52 Jacobs, Tonya L., Elissa S. Epel, Jue Lin, Elizabeth H. Blackburn, Owen M. Wolkowitz, David A. Bridwell, Anthony P. Zanesco, Stephen R. Aichele, Baljinder K. Sahdra, Katherine A. Maclean, Brandon G. King, Phillip R. Shaver, Erika L. Rosenberg, Emilio Ferrer, B. Alan Wallace, and Clifford D. Saron. "Intensive Meditation Training, Immune Cell Telomerase Activity, and Psychological Mediators." *Psychoneuroendocrinology* 36 (2011): 664–81. https://www.ncbi.nlm.nih.gov/pubmed/21035949.

53 Sudsuang, Ratree, Vilai Chentanez, and Kongdej Veluvan. "Effect of Buddhist Meditation on Serum Cortisol and Total Protein Levels, Blood Pressure, Pulse Rate, Lung Volume and Reaction Time." *Physiology & Behavior* 50, no. 3 (1991): 543–48. http://www.sciencedirect.com/science/article/pii/003193849190543W.

condition, so if there's something as simple as meditation that could help manage it, then why not try it? Am I right?

Meditation is not playing the piano, or mowing the lawn, or knitting, or cooking, or running. Those activities, along with sleeping, eating, and exercising, can be done with mindfulness (and will be much better if they are done mindfully), but mindfulness is not meditation. Mindfulness is being aware and conscious of something, being fully engaged in the process at hand—full, real-time presence. Meditation is sitting with the breath and working with the mind and NOT DOING ANYTHING ELSE. Got it? Or do I need to get out my cat o' nine tails again?!

"Meditation brings wisdom; lack of meditation leaves ignorance. Know well what leads you forward and what holds you back, and choose the path that leads to wisdom"

—Buddha

RELATIONSHIPS: THE SHIFTS THAT OCCUR AS A RESULT OF YOUR HABIT EVOLUTION

Relationship with Others Tips:

Get out your calendar and send out calendar invites to people that you want to have dates with. I do this one-on-one with my kids and my friends!

As you go down this path of self-evolution, you are changing your identity. I know, I know. You're probably thinking, *What a load of old bollocks*. But, really, you are. Why? Because about 45 percent of what we do each day are our habits. In a group of studies by Quinn, Wood, and colleagues in which people diarized their daily behaviors, approximately 45 percent of everyday behaviors—such as exercising and eating fast food—were repeated almost every day in the same location.[54]

We know how bloody hard it can be to change our habits, and the neural evidence supports our struggle![55] When we repeat action sequences, sequences of responses get chunked and integrated in our memory with the cues that predict them, i.e., that trigger the whole habit sequence of cue— action—reward. Our brains also do

54 Neal, David T., Wendy Wood, and Jeffrey M. Quinn. "Habits—A Repeat Performance." *Current Directions in Psychological Science* 15, no. 4 (2006): 198–202. https://dornsife.usc.edu/assets/sites/208/docs/Neal.Wood.Quinn.2006.pdf.
55 Barnes, Terra D., Yasuo Kubota, Dan Hu, Dezhe Z. Jin, and Ann M. Graybiel. "Activity of Striatal Neurons Reflects Dynamic Encoding and Recoding of Procedural Memories." *Nature* 437, no. 7062 (2005): 1158-161. http://www.nature.com/nature/journal/v437/n7062/full/nature04053.html.

this, so we don't have to waste energy on thinking about all the different steps—our brains are so unbelievably cool the way they help us out, right?!

What you do every day adds up to how you live your weeks, your months, your years—your entire life.

Also: What you think shapes what you say, what you say shapes what you do, and what you do shapes who you are. And, as I noted earlier, the reason why you do what you do shapes how you feel (remember, if you hold a pencil between your lips, you start to feel happier?!). This play between psyche (mind) and soma (body) means that you change. So, as you shift your daily habits and how you show up on a daily basis, you're shifting your identity. When you shift your identity the people in your life will notice. They will see the changes and, in my experience, their reaction generally goes in two ways.

Some people see the positive changes and want to make them, too, or at least support you on your journey, whereas others are threatened by the fact that you are now unpredictable, and are made aware of their own habits and choices. Making these changes to sleeping, eating, and exercising will require you to shift who you are as a person. When your habits change, your identity and sense of self changes, too.

YOUR EVOLUTIONARY JOURNEY AND THE ENABLER

If the people around you are the kind of people who are enablers—saying, "Oh go on, have a slice of cake," or "Drink a second cocktail with me," then it's going to be much harder to make these changes and it's potentially going to create some arguments. I find that people who do this after you've made a comment about how you are making a change, e.g., cutting out sugar or booze, to be even more annoying and toxic, or is that just me?!

Some people will not appreciate the changes you are making. Your relationships may evolve as you turn to earlier evenings, exercise, or a focus on nourishment, which often lead to a shift in your conversations toward deeper, more connected topics.

So, be aware of this issue, and plan for times to have conversations with those people who play significant roles in your life. Tell them what you're doing, and why it's important to you. My ex-husband, for example, was annoyed that I stopped wanting to stay up late and drink wine and whiskey and eat meat. Instead, I wanted to go to bed earlier, wake up earlier, and eat mainly plants. This difference in how we coped with the stresses of life, and the major shifts that I started to make, contributed to the struggles we were having as a couple.

As I made these changes, I became a different person, one who valued herself much more highly than she had done before. This made it hard for my ex-husband because I was not the same, and I stopped being someone he could control or box in. As I cleared, balanced, and nourished my mind and body by prioritizing my relationship with myself (which had taken a backseat for many years), I started to place a high value on myself. I started to rise up. I became someone who wasn't going to be told what to do anymore, and this was disconcerting and confusing for him. It engendered anger and resentment.

Be mindful, be open, and for goodness' sake—just keep talking about it, right?

YOUR EVOLUTIONARY JOURNEY AND THE SUPPORTER

Some people in your life will notice the changes—abundance, radiant energy, shining eyes, clear skin, positivity, a healthy mind-body relationship—and think, "I'll have what she's having." (That's a *When Harry Met Sally* reference, in case you didn't catch it!) If you create the time and space for connection and deep conversation around your evolutionary shifts, then your friends, family, and partner might start to join you on the path—particularly if you look happier, shine forth an energy that is strong and clear, and demonstrate clear benefits from these changes that you've made.

One of the most important aspects of these shifts is creating the time to have the right kind of conversation, so you can see how the people in your life can support you as you evolve. Understanding Conversational Intelligence—the different kinds of conversation and then how to create the space for the right kind of conversation—can be really helpful.

CONVERSATIONAL INTELLIGENCE

> Create connection. If you work from home, get together a group of women who can co-work together. If you're a parent of young kids, create playdates, do exchanges where you each take each other's kids for time on your own.

When you talk to other people, pay attention to how you talk. Author and organizational anthropologist Judith Glaser studies Conversational Intelligence and talks about three different levels of conversations.[56]

The first is **transactional**—we exchange information. The second is **promotional**—we promote a certain idea and work on convincing the other party of its validity. The third is **transformational**—we create a nonjudgmental space of comfort and we co-create. When we talk, we are navigating back and forth between trust and distrust, between the more-evolved frontal cortex (where we are cultivating empathy and connection) and the limbic system, and the cortisol-producing and fear part of the brain. This is where the magic happens: when we can keep ourselves in the empathic trust part of the brain!

But, as a busy mom of three, I often find that I am more focused on the first and second types when I am parenting, and on the third when I am coaching or in a deep and meaningful discussion with a friend. It's hard when you're parenting or very busy getting things done, to find the time to have the pure transformational conversation and instead get bogged down in the transactional and promotional conversations.

But, the transformational level is where the deep connection happens. In fact, you could say that the transformational level is simply a different kind of conversation—it's a conversation where we are listening to connect, not to judge or evaluate. That can be SO much harder in some settings and requires a real commitment to NOT let the other kinds of conversations become interlopers!

"We have two ears and one mouth so that we can listen twice as much as we speak."

—Epictetus

56 Glaser, Judith. *Conversational Intelligence: How Great Leaders Build Trust and Get Extraordinary Results*. New York, NY: Bibliomotion, 2014.

HOW TO IMPROVE YOUR CHANCES OF CULTIVATING A SUPPORTER

Schedule regular conversations a *deux*. Check in, have time together—*a deux*—that is, the two of you—describe why you are making these changes, and describe the positive effects they are bringing to your life. Make sure you check in once a month or so.

- *Timing is everything*. Having a discussion with someone at 10:00 p.m. about the fact that they denigrate your life choices will not result in good sleep. Schedule a walk on Sunday afternoon or tea on Monday morning before your kids get up to have this conversation.
- *Reactive vs. intentional*. Try not to have this conversation in response to an argument. I had to work on this—quite a few people in my life are avid meat and dairy consumers and will repeatedly make comments about how I am not normal, I am weird, etc., for basing my meals around plants. When I react and get angry in that moment, it doesn't resolve the issue. Having intentional conversations, when everyone is calm, works much better!
- *Explain your big why*. Telling people why this is important to you, why you care about it, and what results you are finding as you make these shifts will help to frame the evolution. For example, I have more stable energy when I avoid animal products, main-lining caffeine, and lots of white carbs/sugars!

AVOIDING TOXIC PEOPLE

You will probably find that as you grow and evolve along this path, the people you want to spend time with might change. I definitely started to find that I did not want to hang out with people who complained about their lives but did not do anything to change their lives or to change the way they thought about their lives.

Jim Rohn, an entrepreneur, author, and motivational speaker, famously said, "You are the average of the five people you spend the most time with," which is very appropriate here. As you make these changes, you may find that your friendship circles are changing and shifting. So, you want to spend more time with the people who have the habits that you are cultivating, who are engaging with the world using ethical frameworks that you aspire to or already use, and who have the mindsets that you recognize as valuable. Some people will start to feel more and more toxic to you, so find ways to limit your exposure to these people, or at least limit their ability to affect you negatively.

CREATING BOUNDARIES

If you cannot avoid these people, because maybe you're related to them or they employ you, then start to think about boundaries. My favorite definition of boundaries comes from author Brené Brown, which is simply: *What's okay and what's not okay*. This is something that will inevitably shift and ebb as relationships shift and ebb, but be mindful of the old adage that we train people how to treat us.

"Not everything that is faced can be changed, but nothing can be changed until it is faced."

—James Baldwin

The key here is not to be a victim. It's really easy to feel like the "world" or particular people are against you, especially when life seems to continually give you lemons. In my view, the real key here is to create your own definitions of yourself—do not let others define you. Do not get stuck thinking that you "ought" to look, think, like, believe (fill in the blank). I recently came across this quotation which I am working on completely embodying "It's not my business what other people think of me."

Step back and acknowledge your strengths and weaknesses and live your life on your terms. When you allow someone else's negative views of yourself to define who you are and how you show up, it can be really easy to let those toxic people take charge of you. One of the ways that I like to think about this is the value of my mind. I consider my mind to be like primo Manhattan real estate. I refuse to let negative, toxic people and negative self-talk become squatters in my valuable real estate!

In the United Kingdom, where I am from, a squatter is someone who lives rent-free in someone else's place—they often break into an unoccupied building and take up residency. I believe we have the same definition here in the United States. As a mother of three, and as a busy entrepreneur, I certainly don't plan on hosting any squatters—not in my home and definitely not in my mind-space!

One of my coaching friends has a list of her non-negotiables that she commits to on a daily basis. She has created a boundary around these things and makes her decisions, schedules her time, and reacts to demands for her time and attention based around these non-negotiables. One of my clients describes these boundaries as fences that he erects to make sure he does what he needs to do and doesn't get drawn into the work of others.

If it's hard to rid yourself of these toxic people in your life, then downscale the kind of contact you have with them, in terms of the emotional valence. My friend and fellow coach Danielle explains it thusly: shift from in-person contact (most emotionally valent) to phone calls or shift from phone calls to emails and/or texts (the least emotionally valent). Something to ponder as you delve into this deep and juicy self-work!

Mindset:

Write down three things you want to achieve in the next three months. What actions do you need to do, to make these happen? Why are these things important to you?

1. Write down what do you every day, from the mundane (bathroom, dishes, pay bills) to the big (yoga, school, run, manage your business).

2. Categorize these into things you love, tolerate, hate.

3. Add up the total and create percentages for love, tolerate, hate. See where you are spending most of your time.

4. Then (1) can you get rid of or delegate the hate events? Or (2) can you change the mindset of things you have to do?

MINDSET

This is essentially about how you think about yourself and others and your skills. As I have mentioned a few times already, how you think changes how you act and feel, and how you act changes how you think and feel, and how you feel changes how you act—and around and around. My point is, these concepts are intimately intertwined, so it's FUNDAMENTAL to your success that you truly and deeply BELIEVE that you can change!

In my view, three things are needed for this to happen. The first is to understand that you can change—the concept of growth rather than fixed mindset. The second is the importance of cultivating gratitude! It's hard to keep on plugging away at life if you don't feel happy and grateful. And the third is the concept of dazzlement!

MINDSET: FIXED OR GROWTH?

Evolution, change, improving your experience in this world, right now—these things are my jam. Remember, it is important for you to get clarity on how you think, feel, and act. Broken record, I know. Although you can't always change what happens to you in your life, what you do have control over is how you FEEL about it. This is mindset. When something challenging happens, do you recoil and run away and give up? Do you swear, cry, pick yourself up, and dust yourself off? Or do you just let it roll off your back?

Mindset is something over which you have power. But first, some terms.

Fixed mindset is the notion that you have a certain set of skills or abilities that you are born with. Part of thinking like this allows you to step back and quit—for example, "Well, I'm not musical, so I won't ever be able to master the piano" or "I don't have a mathematical mind, so I'll never be able to pass the exams to get into medical school." Not to deny that we all have natural proclivities, but when you read the research on the people who meet the bar for excellence in their field, the consistent pattern you find is that they practiced and practiced and, well, practiced some more. So, whatever it is that you tell yourself (or perhaps more worryingly, what you find yourself telling your friends, partner, and kids about themselves and their abilities) becomes this self-fulfilling prophecy. This is a thought pattern that, more often than not, seems to be framed in the negative, which is not going to get you anywhere in life!

Growth mindset is the notion that you can improve the skills and abilities that you were born with. This allows you to envision a future self with skills and abilities that you don't currently have! This is major work for some of us to get our heads around. But, if you can work on it for yourself and with your relationships with others, you will have such a spectacular life! John Assaraf, the spiritual entrepreneur, philanthropist, teacher, and creator of Neurogym, makes a distinction which I think is relevant here. He says that making goals and achieving goals are two different things. If we make goals based on what we have achieved in the past—our education, our (fill in the blank)—then we are deeply limiting ourselves. We need to make goals and then reverse engineer them to get there, by figuring out what we need to learn, what books we need to read, who we need to hire to make this happen.

LEARNED HELPLESSNESS OR LEARNED OPTIMISM?

Historically, psychological research suggested that we do not have a lot of control over the way we feel and act. Author Martin Seligman points this out in his book *Learned Optimism: How to Change Your Mind and Your Life.*[57] For example, he notes that we act as a result of our unresolved childhood issues (Freud), or our genes that direct fixed action patterns (ethologists), or our behaviors are driven by external reinforcement (Skinnerians!), or our actions are a result of our need to reduce biological drives and needs (biobehaviorists).

These different branches of psychology left out the choice of our actions for a large part and put a heavy emphasis on things that were somewhat outside of our control (our genes and our environment). Now we know that what we do affects our genes (epigenetics)—for example, meditation—and we have much more control over our environments now than we used to have. Particularly, I do, as a woman in the twenty-first century.

Research since the 1960s has shifted more towards the control we do have over how we feel and act, and one area has zoned in on helplessness and decision-making. Optimists—glass-half-full folks—believe that when the proverbial hits the fan, they must work harder to resolve the issue. They have a cognitive strategy of telling themselves that what went wrong was not their fault. Confronted with defeat, they try harder. Pessimists, on the other hand, use the learned helplessness strategy in which nothing you choose to do affects what happens to you.

Optimism is a choice. Psychological research in the last half-century has revealed what Buddhists and yogis have known for hundreds of years—that we can choose how we think. The Yoga Sutras of Patanjali lay out the practices for living a yogic life.[58] The four books contain short pithy instructions for us to contemplate and practice. My favorite sutra that relates to this concept of optimism as a learned cognitive strategy states that:

"By cultivating attitudes of friendliness toward the happy, compassion for the unhappy, delight in the virtuous, and disregard toward the wicked, the mind-stuff retains its undisturbed calmness."

—Patanjali, Yoga Sutras I.33.

57 Seligman, Martin. *Learned Optimism: How to Change Your Mind and Your Life.* New York, NY: Alfred Knopf, 2007.
58 Satchidananda, Sri. The Yoga Sutras of Patanjali: the Meaning of the Sutras. San Bernardino, CA: Pacific Publishing Studio, 2015.

The five *niyamas* are the second of the eight limbs of yoga. With the five *yamas* (first limb), they comprise a list of ten suggestions, similar to the Ten Commandments in Christianity and Judaism or the Ten Virtues in Buddhism. *Niyamas* are observances.

Translator and commentator Sri Swami Satchidananda notes, "Contentment means just to be as we are without going to outside things for happiness. If something comes, we let it come. If not, it doesn't matter. Contentment means neither to like nor to dislike."

So, get to it—choose a better model.

And if it feels like you are stuck, remember this Buddhist analogy that I constantly remind my kids about:

The mind is like the sky. Thoughts are like clouds that move across the sky. They are not the sky and they are not permanent. You are not your thoughts and emotions.

I extend this analogy and talk about how some of the clouds have thunder and lightning and others are light and airy. And the key is that they are IMPERMANENT and you can let them go. So, when it feels overwhelming, when it feels that everything is going wrong, remember, you can choose how you feel, and it won't last forever.

"I am not what has happened to me. I am what I choose to become."

—Carl Jung

Learning to increase your perceived control over your world improves your life immeasurably. As noted in the research above, meditation improves perceived control. In the famous Whitehall studies by Michael Marmot, in which he studied the health of the British Civil Servants, he found that perceived control over one's life mediates the effect of stress.[59] Marmot noted that when people climbed the ranks, their stress levels grew (which can increase the likelihood of heart attack, etc.). However, your perceived control increases as you climb the ladder and this has a protective effect. Here it is again—psyche and soma. It's all connected, baby!

59 Marmot, M.G., et al. "Health inequalities among British civil servants: the Whitehall II study." *The Lancet* 337, no. 8754 (1991): 1387–1393.

GRATITUDE

This, my friends, is a biggie. Such a biggie that I have the word tattooed on my inner right wrist to remind me to be GRATEFUL!!

> **Relationship with Self Tips:**
>
> Keep a gratitude journal—what three things are you grateful for today?

Research shows that cultivating gratitude—such as writing thank-you notes, complimenting people, or having a daily gratitude diary to log three to five things that you are grateful for—changes how you feel and changes how you engage with others.[60] For example, your roommates or housemates may report you being nicer if you have a gratitude diary. You may report feeling happier about life if you have a gratitude diary.

What does this actually mean? It looks like this. Every day, take a moment to write down or say out loud to someone (maybe at family dinner time) what you are grateful for. This practice is important because it encourages you to engage in the world in a different way, and allows you to start to look at things, notice them, and feel grateful during your day. You then get another boost when you write it down. And, if you read it again, you get a third boost. ONE event repays you THREE times in feel-good hormones! How freaking cool is that?!

You can vary it, too. For example, I love author and yogi Yeah-Dave Romanelli's "Find something beautiful, tasty, and funny," and, at supper with my kids, I'll say what was beautiful today, what was tasty today, and what was funny today. The key is not to repeat the same thing every day, because then you stop looking and you don't have that extra level of dopamine deliciousness.

Another way to get this going is by doing acts of service. This includes volunteering—connecting yourself to the local community by serving it in some way, such as helping out at the soup kitchen or working with the animals in the animal shelter. My kids and I deliver food to seniors below the poverty line, once a month. Or try something more random and fun—Random Acts of Kindness, in which you do something that creates joy in others (and in yourself), by buying the coffee for the person behind you in the coffee shop, for example, or knocking on doors and giving people unexpected gifts.

---

60 "In Praise of Gratitude." Harvard Health Publications. November 2011. http://www.health.harvard.edu/newsletter_article/in-praise-of-gratitude.

DAZZLEMENT

> Join an online group with similar interests, or with similar challenges or life experiences. It feels sooooo good to have a supportive community around you!

I love this word. Really. I LOVE IT. I once watched a film with a friend, who wanted me to see the film because of the way it documented ritual and he knew my work and love of habit, routine, and rituals! But what I didn't expect when I watched this film was to discover my new favorite word—dazzlement. This word means the state or condition of being dazzled or the action of dazzling! Fantastic, isn't it? So, how can you cultivate this? One way is to make sure you have creativity in your life.

Finding joy, awe, and creativity in your life is hugely important and often ignored. Like play—we often think that play is something that is exclusively for little kids, but actually, play is what allows you to have a more joyful life! And who doesn't want more of that feeling of flow—when you are so embedded in the creative endeavor that you lose track of time?

I was so excited when I discovered that term "flow," coined by Mihaly Csikszentmihalyi (pronounced Me-high Cheek-sent-me-high) while I was doing postdoctoral research as part of the Center for Inquiry in Science Teaching and Learning in the Department of Education at Washington University in St. Louis, MO.[61] One of my roles was to develop programming for kindergarten through second grade in some of the struggling St. Louis public schools by bringing Science Center and Museum educators into the classroom. We co-developed programs to cultivate that excitement and to cultivate that excitement and flow for the students with material that was often missing in the formal classroom (particularly with the focus on teaching to the test). That term, flow, so perfectly described that state of complete engagement in something, which I would see as a docent and facilitator at the St. Louis Science Center (but less frequently in the classroom).

The two elements here involve looking for the awe in your world, which is related to gratitude, and not being focused on the outcome or whether you are any good at it. Einstein reportedly used to go and play the violin when he was stuck on a problem—it gave him that experience of flow, of forward momentum, while giving him

61 Csikszentmihalyi, Mihaly. Flow: *The Psychology of Optimal Experience*. New York, NY: Harper Perennial, 2008.

a break from the task in hand. Can you slow down, observe your world, and make the time and space to create the energy to allow dazzlement to flow?

Creating time and space for awe can seem tough to prioritize in our busy days. Awe, defined as reverential respect mixed with wonder (or fear), is easier to access if you are present. When you are present, your mind and body are in the same place (a lot of the time our minds are not in our bodies—they are wandering off down that Facebook cat-video path, or reliving an experience, or thinking about that dinner we have to go to tonight), your five senses will be attuned to what's going on right here, right now, and tapping into awe is easier.

As I have said many times throughout this book, know thyself. Most of us are visual, so take time to see—really see. I live in Cleveland, Ohio, which has a spectacular, free art museum—the art and the building are awe-inspiring (if I am not on my phone or trying to pacify an argument between my kids)!

I have recently tapped back into an old awe-inspiring activity of mine—playing the piano (I had weekly lessons from ages eight to eighteen). I will sit with a stack of music next to me and play a nocturne by Chopin, a couple of the songs from Les Misérables (I LOVE show tunes!), and then turn my hands toward some ragtime and then back to Mozart. The music rushes over me. I have to be present, scanning the music, or otherwise I stumble and lose track. I can lose time and realize, now, as an adult, that playing the piano was my first experience of true presence, like the feeling I have when I do yoga—meditation in motion. Nothing else matters, nothing else is there. It's just me and the music.

But, as I mentioned earlier, accountability makes it easier to follow through and complete a task, and I wanted to cultivate awe in a different way. So, perhaps meet a friend and sit somewhere awe-inspiring and sketch. When you sketch, you have to really slow down and look. This creative and awe-inspiring practice may not come easily to you, especially if you are out of practice, but don't let that get in the way—if you enjoy it! Check out your local neighborhood—in mine you can paint pottery, do glass blowing, join a knitting circle, make lip balm and candles, have a music lesson! My creative endeavors these last years had been around producing something that was "useful"—a delicious cake for my daughter's birthday, a set of curtains for my son's room, or a knitted woolly hat for my grandfather.

I had dialed down a lot of my creative work as a teenager when I started to become aware of the fact that I was fully mediocre. Mediocrity was definitely a buzzkill when I looked around the art room and agh, ugh, oooof, compared my pictures, pots, and sculptures to ones on the aisles around me. DUH. The creative experience should have been about my experience, and mine alone, not whether mine was better, worse, or more interesting than the ones next to me!

So, sink into what you want—draw, read books, paint, make bowls, bake cakes, knit scarves, play the piano, plant a garden, learn to dance the tango—and then make it happen. Allow yourself the time and space to indulge in creativity. Slow down and look, smell, touch, taste, hear, and allow the awe to bubble up. Find that flow state. Perhaps for you, it's taking apart the engine of a motorbike, or bird-watching, or building model airplanes. Just give yourself permission to cultivate awe and get into that state of flow.

But, before you think that I am just promoting pure hedonism, I am not. I believe in a life of balanced hedonism and eudaimonia, so we live a life of purpose and pleasure.

SUMMARY

Meditate. Seriously—this is a goddamn non-negotiable if you want to create a life of happiness and freedom. Sit your butt down and do it, NOW!

Reflect on the people in your life and whether they are enablers or supporters. How can you dial up the support? Avoid the enablers?

When things don't work out, notice what you do—do you catastrophize, do you jump to conclusions, or do you blame the other people involved? Create boundaries around your time, what you will and won't tolerate.

Finally, make some time in your life for dazzlement. Suggestions: visit your local art museum(s); take a drawing class; pick up a camera; make a complicated recipe; join a meetup for a new hobby—be it conversational Italian, knitting, or making art out of your nose-pickings. Just DO it now!

WHAT DO YOU NEED TO UNLEASH YOUR POWER?

- A plan. Make time—schedule time on a Saturday to hike, for example (get a babysitter if needed).
- To reflect and then commit—journal, sit quietly, ask yourself: what do I enjoy? What brings me pleasure? Famous entrepreneurs, such as Bill Gates, Mark Cuban, and Oprah follow the five-hour rule—give yourself five hours every week to do things that are not related to your work or primary focus (read books, learn a new language, etc.). Doing this, too, will make your life happier!

TIME OUT

I know, I know, I know. Reflecting on yourself, your relationships, how you talk to yourself, and how you talk to others (or let them talk to you) can stir up a lot of icky self-worth talk.

But remember this: paying attention to yourself—e.g., what turns you on, how you like to spend your time, who you want to spend your time with, what you want to do, and how to do it all—this will make for a joy-filled life of freedom, which is why you're here, right? So. Sit still. Close your eyes and take a few deep breaths, up and down your spine—inhale up, exhale down—and chill the eff out, so you can pay attention.

HASHTAGS #Enabler #Supporter #ConversationalIntelligence #KeepTalking #Telomeres #HabitEvolution #IdentityShift #Empathy #SocioEmotionalIntelligence #TransformationalConversations #PromotionalConversations #TransactionalConversations #EnablerVersusSupporter #Evolution #SelfWork #SelfWorth #Boundaries #Awe #Flow #Dazzlement #FiveHourRule

CHAPTER 8

THE END—OR MAYBE JUST THE BEGINNING!

WHY DID I WRITE THIS BOOK?

Well, I want you to have an amazing life. I want you to create dazzlement in your life, a sense of deep gratification and sparkle! I want you to have the freedom and energy to do what you want, when you want, and how you want to do it. And the way to do that is to create some structure in your life.

I've been working with clients for a number of years, and what I have found is that there are particular ways of doing this self-work. When my clients come to me, they are often struggling with exhaustion or stress. They are feeling overwhelmed, unable to completely turn off, and like they are operating at less than their fullest potential. Perhaps they have stopped prioritizing themselves—due to parenthood or a demanding job. I listen to them deeply and hold space for them to completely share who they are and what's going on.

Then, I get them to describe their days to me, so I can see when they sleep, eat, exercise, and do all they need to do. Then we co-construct some frames that will start to create shifts, e.g., a bedtime, a wake time, a lunch time, an exercise time. For some people, it's clear to them that making a frame around their sleeping is going to create the biggest shift, so we start there. For others, it's about exercising, or shifting their diet away from processed foods, or eating heavy meals too late in the day.

The reason I start here is that creating those frames reduces the choices you have to make every day (which are in the tens of thousands). This then creates space and the ability to think and feel deeply about your relationships. When you create some space, you can start working on—your relationship with yourself and your relationships with others. And indeed, for some of us, it's only when we start to care for ourselves well that we even really start to glimpse that the relationships we have with ourselves and others may not be as happy or fulfilling as they could be!

However, others come to me with relationships that are struggling and difficult, or with a deep sense of environmentally-driven sadness or anger. So, the work starts with their relationship with self (meditation, mindset) and/or with others. That might involve creating time to spend together (such as date nights), or maybe it involves attending a couples or parent-child therapist, planning conversations to have with bosses and/or employees, or discussing exit strategies from toxic relationships, or processing some distressing diagnoses.

The point I am making is that you don't have to start with sleeping, eating, and exercising. Start with what is most compelling to you and the area that you feel the most drawn to getting back on track first (for you, maybe it's relationships). What you will probably find is that when you start creating routines around one area, you will start to desire shifts in the other areas of your life, too.

You came here because you wanted to learn how to unleash your power through developing great habits—now you have some tools. Go at it. Go for it. And don't forget to create the structures you need, and for goodness' sake be gentle with yourself when you fall off the wagon…!

THIS IS WHAT THIS BOOK SAYS

- **Face yourself: Know your habits, routines, and what you want or don't want.** Figure out what motivates you—your big why (maybe it's the stick, rather than the carrot, or maybe it's a big-ass reward).
- **Acknowledge the effect the environment has on you and your experience.** See what you can do to change it—change your job, move out of the city, or delegate some of the chores in your life to someone else!
- **Manage your stress.** You need a little eustress in your life, but overwhelm creates issues and decreases the functioning of your mind and body!
- **Meditate.** This is seriously non-negotiable for a happy life. As the journalist and author Dan Harris points out in his book, *10% Happier: How I Tamed the Voice in My Head, Reduced Stress*

Without Losing My Edge, and Found Self-Help That Actually Works, meditation makes you 10 percent happier![62]

- **Schedule time, support, and the things you need to make it happen.** Set goals and then reverse engineer how you are going to get there. Do not limit your goals based on what you did, whether you "failed" in the past, whether your degree is in the right subject (or even whether you have a degree), or whether you need people to help you get there!
- **Understand that it's hard to have the life you want if you're not well slept, well fed, and well exercised.** Hence, here is a basic plan to get on track:
    - **Sleep:** Every day, go to bed by 10:00 p.m. and wake at dawn. Before you go to bed, learn to chill yourself out—e.g., reduce technology, eat smaller meals at bedtime, and pacify your senses. Got it?
    - **Eat:** Plants, mostly plants (a.k.a. whole foods). Have your biggest meal at lunchtime (between 10:00 a.m. and 2:00 p.m.), and your smallest meal at the end of the day (by 6:00 p.m., if you can). Drink lots of water and herbal tea.
    - **Exercise:** Every day. Know yourself—what do you need to unleash your power? Group classes, music, the same thing every day.
- **Create routines and rhythms around your habits.** This may seem like a restrictive, boring thing to do. However, it gives you freedom. When you're not making so many decisions, you have more mental space!

"Honoring your own boundaries is the clearest message to others to honor them, too."

—Gina Greenlee

- **When you mess up, do not despair.** Shouting at your teenager, drinking too much wine, not getting your butt to the gym for two weeks—notice which things, if you can, knocked you off-track the most, and then get back to it. This will give you information about which habits are most important to you now. Don't denigrate yourself, and don't tell yourself

62 Harris, Dan. *10% Happier: How I Tamed the Voice in My Head, Reduced Stress Without Losing My Edge,* and *Found Self-Help That Actually Works.* New York, NY: Turtleback Books, 2014.

you're a useless person. Just get back on the horse and get moving again. Remember—it's consistency, not intensity, that will make the habit stick. Even if you only do one minute of meditation, or five jumping jacks, etc.—that's better than nothing!

- **Look for the dazzlement in your life.** Draw, cook, knit, take something apart, travel, wander the art museum, or learn Italian. Slow down and smell the roses and learn to step away from it all and dial down "the fear of missing out." Play is vital to a happy, successful life.

"It is necessary, and even vital, to set standards for your life and the people you allow in it."

—Mandy Hale

- Figure out how to be in relationships. **You do not operate in a bubble**. Remember the saying, "No (wo)man is an island." Create those boundaries, create those frames and fences, create those habit sequences, and you will see that by following rituals and routines, you can create freedom and joy!

So…I hear you ask…how do I create the kind of community to help support me and my like-minded peeps with this? Here are some guidelines.

HOW TO CREATE YOUR OWN HABIT CIRCLES

How do you move ahead with making successes in the field of habit-shifting?

This is where the rubber meets the road—going from theory to action!

One way is to hire a coach, like *moi*, who you work with one-on-one. For some people, this is the way to go, because you have someone who you connect with regularly and who challenges you to grow, try out new things, and then report back. Often the awareness that you are going to be meeting with your coach motivates you into action!

A second way to move into action is to be part of a group—similar to the twelve-step process—creating a space where people can share and are motivated to change, because they know their current habits are not serving them well. Creating a HABIT CIRCLE where people come together is a fantastic approach.

Here are some tips for creating your own group—either online or in person.

CREATE YOUR OWN HABIT CIRCLE

1. **Logistics**

    A. Decide on a structure—e.g., once a week or once a month.

    B. Decide on whether to do it in person or online. It's easy to create a Facebook group! Face-to-face connection is deep and meaningful, but can be hard with our busy lives.

    C. Reminder invites.

    D. Location if in person.

2. **Create a Mission or Goal**

    A. What's the point? What are you gathering for?

    B. Are you going to focus on particular habits—are you going to do this, #bosslady or #bossman, or is this going to be a collective decision?

3. **The Role of Organizer/Host**

    A. Your role is to hold space for everyone, guiding conversation.

    B. Make sure everyone feels welcome and heard. Create space for everyone to share—e.g., going around and asking everyone to share.

    C. Create the space for the WHAT and WHY, get behind why people want to change—what's the point?

    D. Help people see themselves being successful—the vision of what they want to become/achieve.

    E. Create space for the HOW to deal with this habit-shift.

    F. Discomfort is part of evolution—why? Because you are stepping into the unknown and your brain is primed to keep you safe, and the unknown feels unsafe! Remind participants that this is part of the journey—working through the unknown and the fear.

4. Managing Participants' Interactions

A. Avoid prescriptive advice.

B. Make it clear that everyone in the group is required to engage. We all impact each other and everyone needs to take part if there's going to be transformation. Prod the lurkers!

C. Remind everyone to have an open mind—the Beginner's Mind. This is a notion that, as we get better at something, we think there is only one "right" way to do something. Remind everyone that that's not the case!

"Each time you say yes to a healthy boundary, you say yes to more freedom."

—Nancy Levin

Changing habits when we are in relationships is different from when we are just operating on our own. We are operating in the context of others. What became epically clear to me during a traumatic six-month period of late 2017 through early 2018—when I was hacked and stalked, and then I herniated a disc in my back and was once again struggling with these events which had shaken my cage—was the role of community.

Who were the people in my life who were there for me, when I was deeply struggling? How did I show up in these different communities?

There is a Buddhist word, *Sangha*, which is traditionally used to refer to groups of monks or nuns who cohabit and share living spaces, teachers, and so on. I think of all the sanghas that I operate within and the habits these different communities have, how that impacts me and my life, and how easy or difficult it is to change my habits or to continue my good habits within these communities.

Join me in a life of spectacular freedom and epic awesomeness—it's in your reach!

TAMSIN'S RECIPES

PLANT FUEL FOR A HIGH VIBE MIND & BODY

TABLE OF CONTENTS

Tamsin's Top Kitchen Tips

In Ayurveda we have six tastes: Sour, Sweet, Salty, Pungent, Astringent, Bitter. Having each of those tastes in every dish helps with the feeling of being satisfied. If you have done your dosha tests, you will know what your predominant dosha is and thus which foods to avoid or to eat.

However, I don't want you to get too hung up on this, especially if eating a plant-based diet is new to you.

My key tips are:

1. Think more about WHEN you eat (remember, biggest meal at lunch);

2. Think about the SEASON you're in (warmer foods in winter, bitter foods as you come into spring, cooling foods in the summer); and

3. PREP, PREP, PREP, and—oh, did I mention? PREP THE ELF OUT OF YOUR KITCHEN!

    A. Marinate beans/lentils and bulk cook in rice cooker or instapot for the week, add a bouillon cube, a tablespoon of diced garlic, spices—whatever floats your boat! Use these to add to salads, to soups, to serve over grains.

    B. Roast a tray of veggies, etc. (rutabaga, squash, turnips, carrots, onions, sweet potatoes, whole head of garlic, top sliced off and wrapped in foil).

    C. Chopped-up veggies and hummus or variations for you and the kids for snacks.

    D. Melon balls, grapes—cut into easy-to-grab handfuls. You're MOST likely to eat what is at eye level and is prepped, so make it easy on yourself and your kids.

    E. Diced veggies (tomatoes, grated beets, red peppers) and herbs ready to sprinkle on the top of your soups and chilis.

4. Base your meals around VEGETABLES. You know—eggplant and asparagus, rather than steak and pasta.

5. Buy organic/non-GMO/local wherever possible.

In 2018, the dirty dozen (highest dose of pesticides) are: spinach, strawberries, nectarines, cherries, celery, potato, apples, grapes, peaches, pears, tomatoes, bell peppers.

Check the Environmental Working Group site (EWG.org) or download their app, Healthy Living (includes products for home, body, etc., too) for the "dirty" and "clean" fruits and veggies each year, according to pesticides, and try to buy organic.

Tongue Teasers

The key to exploring food that is plant-based is to get comfortable with condiments and spices which will help your taste buds dance!

Have jars of vinaigrettes—lemon juice and olive oil, or Apple Cider Vinegar (ACV) and olive oil (equal amounts)—add grated ginger and garlic or ½ tsp. of Dijon mustard. Use hummus instead of oil. Stir in some Vegenaise.

I like to have on hand:

- Sriracha
- Frank's RedHot sauce
- Vegenaise
- Soy sauce
- Mustard—e.g., Dijon
- Vegan Worcestershire sauce
- ACV
- Flax oil
- Extra Virgin Olive Oil (EVOO)
- I also like to have chutneys, sauerkraut, and beet-kraut on hand to add to my dishes!!

BASIC VINAIGRETTE

- Juice of 1 lemon or lime
- Equal amount of EVOO
- Seasoning

Variations

Add Vegenaise, hummus—replace oil, ½ tsp. Dijon mustard, 1 tsp. fresh ginger and 1 tsp. garlic, splash of Sriracha.

Start having fun—food is SO fricking fun if you just let yourself explore the colors, flavors, smells and so on!!

SPICES AND HERBS FOR YOUR CUPBOARD/FRIDGE

- DRY: hing (asafoetida), cumin, fennel, coriander, cinnamon, cardamom, nutmeg, turmeric, paprika, curry blends, e.g., garam masala, chili powder
- FRESH: ginger, garlic, cilantro, parsley, thyme, and we all know about basil!

OILS/GHEE—OR USE VEGGIE BROTH/WATER TO SAUTÉ

- For sautéing use HIGH smoke points, e.g., coconut, peanut, avocado, ghee (clarified butter—no lactose), butter (protein and sugar which can smoke)
- For drizzling after cooking/on raw food you can use LOW smoke points, e.g., olive oil, nut oils, flax oil

DRIED FRUITS/NUTS/SEEDS TO ADD TO YOUR GRANOLA, OVERNIGHT OATS, TO YOUR SALADS

- Blueberries, mangoes, raisins, cranberries
- Pine nuts, sliced almonds, crushed walnuts or pecans
- Sesame seeds, sunflower seeds, chia seeds, flax seeds

Breakfast

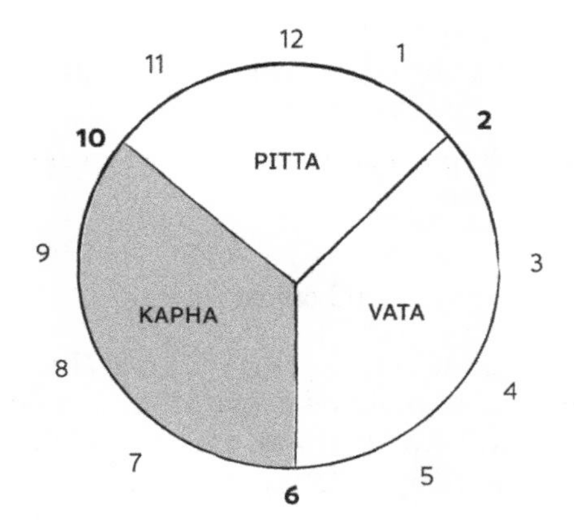

6–10 a.m.
10–2 p.m.
4–6 p.m.

Most days, I presume you will be eating this before 10 a.m., so you will most likely be eating in KAPHA time (6–10 a.m.) which is not great for digesting heavy meals. So—these are nutrient-dense and not too hard to digest. Your first meal of the day, when you break your fast, sets the tone—so make it good!

SWEET

Some mornings you wake up craving the sweet and tasty on your tongue. Here are some delicious ways to begin!

GREEN SMOOTHIE

Yields 1 serving

Ingredients

- 2 c. water (or a mix of: water, nut milk, coconut water)
- Big handful greens (spinach, kale, parsley, lettuce etc.)
- 1 sweet thing (banana, apple, pear ½ c. apple sauce, 1 tsp. honey/agave syrup)
- Optional sprinkle of cinnamon, cardamom, nutmeg, turmeric

- Optional handful of dried fruit (raisins, goji berries, cranberries, etc.)—avoid if blood sugar is high
- Optional handful of nuts (eating nuts 7 days/week reduces chance of death by 20 percent from heart disease and cancer!) or 2 tbs. nut butter
- Optional tablespoon of seeds (chia and flax—omega-3s)

Method

1. Blend until smooth and enjoy!
2. If your smoothie is lumpy, soak the nuts, seeds, and dried fruit in 1 cup of water overnight in the blender. Nuts and seeds will give you more energy to last through the morning. Warm up milk/water in cold months.

WARM GRAINS—SUCH AS OATS, BROWN RICE, QUINOA, OR SORGHUM

Yields 1 serving

Ingredients

- ½ c. of your grain (oats, barley, quinoa, sorghum, rice)
- Fruit (1–2 tbs. dried fruit: raisins, cranberries, dates, ½ c. frozen berries, ½ chopped fresh banana)
- 2 tbs. chopped nuts (cashews, almonds, walnuts, pecans) or 1 tbs. chia seeds—for some added protein
- 1 c. water
- Optional apple sauce

Method

Put all the ingredients (unless your fruit is BANANA—in which case, add after) in your saucepan or rice cooker and cook. If your dominant dosha is Vata (air/ether) and/or it's fall/winter, I suggest putting the dried fruit and nuts in the saucepan with the water (or ½ c. of apple sauce) to hydrate them, which makes them more digestible. But, the crunch is YUM, isn't it?!

BAKED APPLE

Yields 1 serving

You can batch cook a whole load on a Sunday afternoon and have them for the week...! Just sayin'.

Ingredients

- 1 apple
- 1 tbs. nuts
- ¼ *tsp.* cinnamon (cardamom, nutmeg optional)
- ½ tbs. butter/coconut oil/ghee
- 1 tbs. raisins or ½ tbs. raw sugar or alternative—see below

Note: If I'm using a sour apple, like Granny Smith, I like to include the raisins or sweetener. If I'm using a sweet apple, like Jazz or Braeburn, I reduce or avoid the raisins and sugar because it's too sweet.

Method

1. Preheat oven to 350 degrees F.
2. De-core apple, LEAVING A BASE, and scoop out middle. Mix all ingredients and stuff gently inside the apple.
3. Bake for about 30 minutes.
4. Delicious served with a yogurt or coconut cream, or mashed and stirred into a bowl of simple warm grains.

COCONUT CREAM

Ingredients

- 1 x 13.5 ounce can full-fat coconut milk (refrigerated overnight)

Method

1. Turn the can of coconut milk upside down and open top with a can opener. Pour the coconut water into another container.

2. Scoop the thick coconut cream out into a large bowl.
3. Beat cream with a hand beater until soft peaks form.
4. Transfer to a container. Whipped cream can be stored covered in the refrigerator for up to one week.

OVERNIGHT YOUR OATS SO YOU DON'T GO POSTAL!

I get sooo hangry sometimes, don't you? Like, give me the damn food now or I am going to kill someone. This was super strong when I was breastfeeding—the hunger was insane. But I have also found that when I work out, eat my last meal of the day early, and eat plants, I get really deeply hungry, in a good way. These are great to have on hand. I make them en masse so they last for a few days.

Yields 1 serving

Ingredients

- ½ c. oats
- ½ c. milk (I like almond, coconut, or oat milk)—or ¼ c. apple sauce, ¼ c. milk
- 2 tbs. chia seeds and/or 2 tbs. chopped nuts

Now—have some fun:

- Add mashed banana, apple sauce
- Add 1–2 tbs. of nut butter
- ½ tsp. pumpkin pie spice, or just cinnamon

Method

Mix ingredients in a glass Tupperware container or Mason jar. Refrigerate. Eat for breakfast: cold, or warm! Add fresh berries.

GRANOLA

Yields 6–8 servings

Ingredients

Dry Ingredients

- 2½ cups rolled oats
- ½ c. sunflower seeds (or pepitas, flax, sesame, or replace half of it, i.e., ¼ c., with unsweetened coconut)
- 1 cup chopped nuts (e.g., almonds, walnuts, pecans)
- ¼ cup cooked millet (or buckwheat groats or more oats)
- ½ tsp. ground cinnamon
- ½ tsp. nutmeg/cardamom/cinnamon or pumpkin pie spice
- ½ tsp. salt

Wet Ingredients

- ½ cup maple syrup
- ¼ cup oil—I like using coconut oil (solid at 70 degrees F, so warm it)
- 1½ tsp. vanilla extract
- 2 tbs. blackstrap molasses (optional)
- Optional tongue teasers: ½ cup dried cranberries—or raisins, dried blueberries

Method

1. Preheat oven to 275 degrees F.

2. Mix dry ingredients in a large bowl—do this properly so spices/salt get evenly distributed!

3. In a glass jug, stir together the wet ingredients. Pour this over the dry mixture and stir until everything looks coated. Add the cinnamon and the salt and stir.

4. **If you like it clumpy,** pour it onto the baking sheet, pack it down evenly, put it in the oven, and leave it for about 20 minutes, then check it and spin it around and cook for another 15 minutes or so. Different ovens have varying warm spots, so I move my granola tray, cakes, etc., around.

5. **If you like it not-clumpy,** pour mix onto baking sheet and put in the oven for 40 minutes. Stir it every 10 minutes so that it cooks evenly.

6. You are looking for the oat-mix to be a nice golden toasted color.

7. Add dried fruit when you pull it out of the oven. LET IT COOL before you store in jars/containers (otherwise it will get soggy)! It looks super cool on the counter-top in large mason jars or flip-top glass containers, you know, like totally Insta-ready!

SAVORY

You know those days when you wake up feeling salty, I mean, like eating some salty, savory foods. If you're used to eating eggs on toast with a side of bacon, try these options out. Here are some tasty ways to break your fast...

OATS

Ingredients

SALTY, 'SHROOMY OATS

- ½ c. oats
- 1 c. water
- ½ c. sliced mushrooms
- Seasoning

For flavor:

- Add 1 tsp. nutritional yeast (Parmesan-like dried flakes high in vitamin B12)
- ½ tsp. hot sauce
- ¼ tsp. Dijon mustard
- Or play around with your favorite spices!

Method

Mix everything together and cook until done!

SALTY, SPINACHY OATS

- ½ c. oats
- 1 c. water
- ½ c. spinach

For flavor:

- Add 1 tsp. nutritional yeast (Parmesan-like dried flakes high in vitamin B12)
- ½ tsp. hot sauce
- ¼ tsp. Dijon mustard
- Or play around with your favorite spices!

Method

1. Mix everything except the spinach together and cook for 5 minutes or so. Add the spinach in small amounts—using a lid helps get the heat high enough to wilt the spinach. Or mix both recipes together! And have Salty, Shroomy, Spinachy Oats!
2. Add diced tomatoes on top!

TOMATO STEAKS

Ingredients

- 1 beef tomato, sliced into thirds horizontally
- 2 handfuls of spinach
- Optional: 1 piece of amaranth and quinoa toast
- Ghee or coconut oil
- ACV or lemon juice
- Optional tongue teaser seasoning for steaks: turmeric, paprika, salt, pepper, thyme, oregano, a spice blend, vegan Worcestershire sauce!
- Fresh parsley and cilantro to sprinkle on top!

My mother used to serve me warmed tinned plum tomatoes on toast when I was a kid. This is a twist on that.

Method

1. Warm up your frying pan. Add a little ghee or oil. Season your tomatoes, if using. On a medium heat, fry your seasoned tomato steaks. Wilt your spinach in a saucepan with a little water and then gently drain out the water.

2. Toast your toast if using.

3. Serve your tomatoes on the toast, with the spinach on the side, or serve your tomatoes straight onto the spinach! Splash some ACV or lemon juice on top!

MUSHROOMS ON TOAST/TOMATO OR SPINACH

Ingredients

- 1 c. mushrooms
- 1 piece of toast
- Optional tongue teasers: parsley, chives, garlic, hot sauce
- Ghee/oil/veggie stock

Method

1. Sauté 1 cup of mushrooms—you can add garlic and fresh parsley or chives.

2. Serve on toast or a bed of steamed spinach, or on a tomato steak with cracked pepper.

3. Add a tomato steak on the side if you fancy!

AVOCADO ON TOAST

Ingredients

- Half an avocado
- Juice of half a lemon
- Seasoning
- Optional tongue teasers—add diced tomatoes, splash of Worcestershire, fresh chopped cilantro, hot sauce, sesame seeds, flax seeds

Avocados are nature's butter! They don't need any oil.

Method

1. Take half an avocado.
2. Scoop it out into a bowl, mix with the juice of half a lemon, some salt and pepper, and optional tongue teasers.
3. Smash onto toast and serve!

Start to play! Have fun with food!

Soups

BASIC FORMULA

- Soups need a base—garlic, onions/leeks/green onions/shallots, celery
- Some (optional) extra flavor: ginger, spices
- Some optional greens (e.g., spinach, kale, chard)
- Something to thicken them—potatoes, sweet potatoes, fennel, sweet corn, squash, zucchini, beans/lentils, bread (use as last resort)
- Broth, 2–4 cups/batch, e.g., low-sodium bouillon cubes

SIMPLE VEGETABLE SOUP

Ingredients

- 1 tbs. oil, or veggie stock or water to go fat-free
- 3 diced shallots (they have a subtlety that regular onions lack!!)
- 1 clove minced garlic
- Peeled sliced fresh ginger (1" cube)
- 1 cubed zucchini or squash (1")
- 1 peeled sweet potato, cubed (1")
- 2 handfuls of coarsely chopped greens (spinach, kale, chard—do more spinach since it reduces)
- 2 veggie stock cubes dissolved in 4 c. water
- 2 bay leaves

- Spices (e.g., ½ teaspoon each of cumin, coriander, and turmeric, and a shake of cayenne or hot paprika, OR ½ tsp. of a blend, e.g., garam masala or a curry mix)
- Optional tongue teasers: sunflower seeds and cilantro

Method

1. Fry shallots, garlic, ginger, and zucchini in a splash of water or oil until they soften.
2. Add spices.
3. Add stock, sweet potatoes, and bay leaves, and cook until the sweet potatoes are softened.
4. Add greens and cook until wilted. Remove bay leaves.
5. Blend and garnish with sprouts, toasted seeds/nuts, fresh herbs. Or don't and just drink it out the blender. Just kidding…

TOOT SOUP!

As my grandmother would say, "Where e'er you may be, let your wind go free," and my other grandmother would blame it on the dog!

Yields 4–6 servings

Ingredients

- 5 leeks
- 2 bay leaves
- 2 15-oz. cans white beans (e.g., Great Northern, cannellini, baby lima beans). Navy beans tend to lose their skin FYI!
- 3 cloves garlic
- 2 handfuls of fresh green herbs (I like half oregano and half parsley—also try basil, marjoram, a little cilantro), coarsely chopped
- A few sprigs of fresh thyme (leaves only), coarsely chopped
- 2 tbs. oil or ghee, or a splash of the veggie stock
- 6½–7 cups water with 3½ veggie bouillon cubes stirred in (you'll need that extra ½ cup if you cook with veggie stock, not oil)

- Optional tongue teasers: ⅓ c. coarsely chopped parsley for serving, hot sauce/chili oil, pesto

Method

- Cut the bottoms off the leeks and slice lengthwise, washing out the mud and grit. Slice into thin rings.
- Heat the oil/ghee/splash of stock on a low temp and add the leeks, thyme, and bay leaves, so they sweat and don't burn, until they are soft, which takes about 10 minutes. If you're using the stock, keep adding as it evaporates.
- Stir in the garlic and sauté for about a minute. Add the stock, beans, and fresh green herbs and season. Simmer for about 20 minutes (different beans have different skins/sizes).
- Take out the bay leaves and serve, stirring in the extra parsley, maybe adding a little pesto and drizzling with hot sauce for a totally gorgeous dish!

ZUCCHINI AND TOMATO SOUP

Ingredients

Yields 3 servings

- 2 tbs. oil
- 2 zucchini
- 2 large tomatoes, chopped (or a pint basket of cherry tomatoes)
- 2 veggie stock cubes dissolved in 4 c. water
- handful of fresh oregano and fresh parsley
- hot sauce (e.g., Frank's Sriracha) or chili oil
- salt and pepper

Method

1. Slice zucchini into ⅓-inch rings and fry in a heavy-bottomed pan or on a griddle, until just done (al dente).
2. Dissolve stock cubes into 4 cups of warm water.
3. Add chopped tomatoes (or cherry tomatoes), grilled zucchini, and chopped fresh herbs, and bring to the boil.

4. Turn off heat. Taste and adjust seasoning.
5. Serve with crusty bread and a drizzle of hot sauce.

ROASTED ACORN SQUASH SOUP

Yields 2 servings

Ingredients

- 1 acorn squash
- 1 onion or equivalent (e.g., 4 or 5 green onions, 3 shallots)
- 1 diced carrot
- 2 cloves finely chopped garlic
- ½ tsp. cinnamon
- 1 tsp. nutmeg
- 1 veggie stock cube dissolved in 2 c. water
- optional: ⅓ c. cream or milk, or alternative
- optional: cilantro to tear up and sprinkle for serving (makes it look totally awesome)

Method

1. Preheat oven to 400 degrees F.
2. Cut the squash in half and scoop out the seeds. Place the squash face down on a baking tray and pour about an inch of water around the squash. Bake it until soft (depends on size—40 minutes to 1 hour 10 minutes or so).
3. While it's baking, chop the onion, carrot, and garlic, then sauté them until soft and fragrant (keep the heat low to keep the garlic from burning).
4. Go and read a book or meditate or go for a run. When the squash is done (toothpick or skewer pierces skin and flesh easily), grab your blender. Scoop out flesh and add the contents of the sauté pan, the stock, and the spices and blend until smooth. Add cream if desired.

I also roast another acorn squash, face up, with a splash of half and half or non-dairy equivalent, or ghee or veggie stock and some cracked pepper and salt and ⅛ tsp. of ground nutmeg, to eat another day. SO tasty!

You can also stir half a squash into your red sauce to add a little extra vitamin C, vitamin A, vitamin B6 and magnesium—upping the functioning of your hawt bod!

Family Dishes—Cook, Serve, and Eat In One Pot

Eat at lunch or smaller amounts at dinner—remember, your evening meal should be LIGHT and SMALL and EARLY!

NO-JOKE MULTIPLE MEAL CHILI

Yields 4–6 servings

Ingredients

- 2 tbs. oil or water
- 1 onion or 3 shallots
- 2 cloves garlic
- 2 peppers (green, red, yellow)
- 1 jalapeno pepper (or a small tin of chopped jalapenos)
- 2 15-oz. cans beans (e.g., one can black beans, one can kidney, if you want to get fancy)
- 1 15-oz. can refried beans (black or pinto). Or, another can of beans, which you smash up with back of fork, to give a thickness to your chili
- 1 15-oz. can chopped tomatoes, or a few fresh tomatoes, coarsely chopped, with their juices and 1 tbs. of tomato paste
- 1 tsp. ground cumin
- 1 tsp. turmeric (this is a mighty spice and has numerous health benefits!)
- salt and pepper

Method

1. Fry onions, peppers, and garlic in vegetable oil, at a medium heat, until the onions and peppers soften.

2. Add cumin and turmeric and stir for a moment—dry cooking spices makes the above ingredients more fragrant.
3. Stir in beans and tomatoes. Add refried (or smashed) beans and cook for about 10 minutes.
4. Taste and add seasoning if needed. Serve with diced tomatoes, chopped green onions, avocados and sour cream and grated cheese (or non-dairy alternatives), over rice, baked potatoes, or a mound of raw or steamed spinach.

No joke. This is simple to make and can be used for several different meals and can be frozen.

Make in bulk. One night eat it on spinach, one night eat it on rice. Roll it up in lettuce leaves. Add 2 cups of bouillon and make a soup and add a mountain of fresh cilantro. Stick in tupperware in the freezer as an easy back up weekday meal to have on hand!

KITCHARI

Mono-diet cleanse, when your system needs a reboot! Or just a tasty dish for a cold day!

Yields about 8 servings

Ingredients

- 2½ c. white rice (if doing a cleanse; otherwise, brown rice is fine)
- 1 c. mung daal (if you're not detoxing, you can use other harder to digest lentils/beans)
- 6 c. water or 5 c. water plus 1 c. coconut milk
- 1 inch ginger, chopped
- Salt and pepper
- 2 tsp. ghee or oil
- 1 chopped green onion or shallot
- Optional: 2 c. chopped veggies—sweet potato, zucchini, carrots
- Spice mix: e.g., ½ tsp. coriander powder, ½ tsp. cumin powder, 1 tsp. turmeric, ½ tsp. mustard seeds, 1 pinch of

asafoetida/hing (optional). If this seems too overwhelming, find a blend that you like, e.g., garam masala or curry

- Optional: cilantro for serving (it's good for detoxing)

Method

1. Remove stones from daal and rice—wash well! Cook rice and daal in water, with salt, covered, until soft, about 10 minutes.
2. Chop veggies and add to rice/daal mix and cook for 10 minutes more.
3. Sauté the spices (releases their flavors)—in oil or dry, low heat, and add to the veg/rice/daal mix.
4. Sprinkle with cilantro and serve!

CLEANSE

Eat this food every day, three times a day for at least three days, up to two weeks, if you're doing a cleanse: vary the vegetables and spices, e.g., coconut milk and nutmeg, cinnamon, cardamom etc.; thyme/parsley/basil/oregano.

VEGGIE NOODLE STIR-FRY

Ingredients

Sauce

- 1 tsp. honey
- 1 tbs. orange juice
- 1 tsp. sesame oil
- 2 tbs. soy sauce
- 2 tbs. peanut oil
- Mix together

Stir-Fry

- 2 oz./person dried rice or egg noodles (rice is good if you're gluten-free)
- 2 finely sliced carrots
- 1 sliced stick celery

- 1 diced pepper
- 1 c. sugar snap peas (or broccoli in small florets, or green peas/beans)
- 1 chicken breast, cut into 1-inch pieces or a block of tofu cut into 1/2 inch cubes
- 2 green onions, a leek or half a regular onion
- Optional ½ c. cashew nuts—I love adding these, reminds me of my local Chinese restaurant Kam Tong in Queensway, West London, where my grandmother would take me as a kid!
- 1 tbs. oil
- Optional tongue teasers: 1 c. bean sprouts, ½ c. baby corn or corn sliced off a cob, handful chopped cilantro for serving.

Method

1. Set a pot of water on stove to boil. Prepare the meat/tofu if using (not necessary). Make cubes bigger than a dime, smaller than a quarter! If using tofu, drain it and press out liquid, by pushing into a sieve or wrapping block and pressing it until most of liquid comes out.

2. Heat 1 tbs. oil and cook tofu or chicken in a big wok/heavy bottomed pan with sides—7–8 minutes until browned and just cooked. Take it out of pan.

3. Use remaining oil and cook carrot, pepper, and onion for 2–3 minutes, then add corn, cashews, peas, and sauce (I use any veggies I have in fridge).

4. Add noodles to boiling water (usually about 4 minutes, but read package).

5. Drain noodles and toss everything together and dress with cilantro!

I say to my kids—use chopsticks and pick up the stuff you like and leave the stuff you don't…

TRI-COLOR OAT RISOTTO

Ingredients

- 2 tbs. ghee or oil
- 2 c. rolled oats
- 4 tsp. chopped fresh sage
- 3–4 c. water
- Handful of fresh spinach
- 10 halved cherry tomatoes
- 1 tsp. sweet paprika (and if you like a kick add ¼–½ tsp. of hot paprika)
- Salt and pepper

You could eat this for your salty breakfast too!

Method

This is like a traditional rice risotto, but the oats cook much faster, are rich in antioxidants, and have been shown to lower cholesterol. If you substituted risotto rice (Arborio), you'll need more time to cook.

1. Heat the oil. Stir in oats and cook for about 3 minutes.
2. Add sage and cook for another 3 minutes.
3. Stir in water, paprika, and seasoning. Cover and reduce heat until water is absorbed and oats are fluffy—less than 10 minutes.
4. Add tomatoes and spinach (if using) about halfway through.

Cold Dishes for a Hot Mama

Eat at lunch or supper, or for breakfast if the mood takes you—break some of those "rules" about what you should eat when, right?!

The key to these raw salads is to let them marinate!

SIMPLE MARINATED SALAD

Ingredients

- A handful of chopped, sturdy leafy greens (e.g., kale, turnip greens, mustard greens—not spinach or lettuce)
- ½ tsp. salt

Salad Dressing

- Juice of ½ to 1 lemon (some lemons are tiny!)
- Equal amount of oil (to lemon juice)
- 1 tsp. finely chopped garlic
- 1 tsp. finely chopped ginger

Any Mix of the Following

- ¼ c. seeds or chopped nuts (e.g., pumpkin seeds, sunflower seeds, chia, hemp, almond, walnut)
- ¼ c. dried berries (e.g., cranberries, blueberries) or raisins
- ½ grated carrot
- ½ c. grated raw beetroot
- ½ avocado
- Chopped tomatoes
- Chopped green onions, cilantro, parsley—as you wish
- Crumbled goats cheese, feta, etc., if you eat cheese!

At Thanksgiving, it's lovely to make this with cranberries and toasted pecans!

Method

Wash your hands, roll up your sleeves, and mix greens with salt to start the breaking-down process. Add carrot and beetroot if using and toss with salad dressing. Leave for at least 30 minutes. Become one with the bunny rabbits that like eating this kind of thing.

You can leave this for a day or two in your fridge, as long as you add the soft ingredients and onions, etc., on serving.

The salt and lemon make the raw ginger and garlic easier to digest by partially breaking them down.

HUMMUS WITH A FEW TWISTS

Ingredients

- 1 15-oz. can chickpeas/garbanzo beans
- 2 cloves crushed garlic
- Juice of ½ to 1 lemon (I like things very lemony, so will go for 1 or more lemons) OR lime
- Optional: 1 tbs. tahini—if you're making it for kids, start without and then add 1 tsp. and build up to 3 tsp. (1 tbs.)
- Dash of olive oil
- ½ tsp. salt
- Optional hot sauce

Method

Blend it all together. Eat with chopped raw veggies or some tortilla chips or some of those yummy gluten-free bean chips!

Variations

- For pink hummus—add 1 cooked beet
- Exchange garbanzo beans for cannellini beans and instead of tahini, add 1–2 tbs. of parsley

Make Life Easy On Yourself

I always have glass Tupperware containers of cooked quinoa, rice, lentils, or beans in my fridge. Chopped veggies, fresh herbs, and vinaigrette fixings. Have prepped fruits that last—e.g., melon balls, cubed pineapple, little bunches of grapes.

Easy Weekday Meals

Build a plate from the following ingredients:

- Greens (e.g., raw spinach, lettuce, or steamed/sautéed kale, chard—maybe with garlic or spices)
- Pre-roasted veggies (e.g., turnip, parsnip, rutabaga, carrot, onions, squash, eggplant)
- ½ c. quinoa
- ½ c. cooked lentils, or the chili (warm in the winter, cold or warm rest of the year), ½ c. quinoa, or the tri-color oat risotto, or kitchari (warm in the winter, cold or warm rest of the year).

And then garnish

- Fresh herbs
- A dollop of hummus/baba ghanoush/white bean dip
- Diced peppers
- Tomatoes
- Sliced avocado
- Sautéed mushrooms
- Lightly toasted seeds or nuts
- Vinaigrette

Boom! You have a healthy meal.

Variations

- OR—take one of the soups, and add some lentils, some quinoa, some roasted veggies and diced prepped spices/veg!

DELECTA-BALLS

Ingredients

- ½ c. raw nuts ,e.g., walnuts, cashews, almonds (or buy pre-ground, e.g. almond meal, which is more expensive, but the blending is already done!)
- ½ c. nut butter, e.g., cashew, almond, or peanut

- 6 dates/prunes or ½ c. soaked raisins
- 3 tbs. dark choc. chips (or unsweetened shredded coconut)
- 3 tbs. rolled oats
- 1 tbs. vanilla
- ½ tsp. cinnamon
- 1–2 tbs. water
- Optional cocoa powder, unsweetened shredded coconut

Method

1. Blend nuts until flour-like.
2. Add the nut butter, dates or hydrated raisins, and oats. Then the liquids—1 tbs. at a time—watch the mix, you don't want it to get too sticky. Then mix in the chocolate chips if using. I have a Vitamix and find that adding the liquids from the start makes it easier.
3. Stick your lovely clean hands into the blender (after unplugging it—doh!) and make little ½–1-inch diameter balls by rolling the mix around your palm.
4. If you're using the cocoa powder, or unsweetened coconut pieces, roll the mix in the cocoa powder/shredded coconut so they look like swanky little truffles and place in a glass jar or Tupperware for the fridge.
5. Or, you can smush them flat so they look more like a traditional cookie. Good for packed lunches too. I think they taste better cool, but frankly when I make them I often just eat a few right then and there!

HOT SPICY FRUIT SALAD

If serving with coconut cream, refrigerate can overnight, or during whole day!

Yields 2 generous servings

Ingredients

- 1 banana or 2 mini bananas
- 1 lime
- 2 4-inch wedges of pineapple
- 2 4-inch wedges of papaya or mango
- 1 tbs. butter/ghee/coconut oil
- 2 tbs. coconut flakes (unsweetened)
- ¼ tsp. powdered ginger
- ¼ tsp. ground nutmeg
- ¼ tsp. ground cumin or seeds
- ¼ tsp. ground coriander or seeds
- Optional tongue teasers: ¼ tsp. cinnamon, ¼ tsp. cardamom
- Crème fraiche, plain/unsweetened yogurt, or whipped coconut cream (my favorite) for serving!
- Swank-appeal: a couple of physalis fruits, with the papery layer peeled back, or 2 wedges of star fruit for serving!

Method

1. Take your 1 13.5-ounce can full-fat coconut milk (refrigerated overnight).

2. Turn the can of coconut milk upside down and open top with a can opener. Pour the coconut water into another container. Scoop the thick coconut cream out into a large bowl. Beat cream with a hand beater until soft peaks form. Transfer to a container. Whipped cream can be stored covered in the refrigerator for up to one week.

3. Prepare fruit. Slice banana in half with the skin on and then lengthwise, so you end up with 4 quarters (it's fun to do this with the mini bananas too). Cut lime in half. Cut a few long wedges of pineapple without prickly outsides (roughly

rectangular, 3–4 inches long by 1 inch). Cut papaya or mango into similar long wedges as pineapple (I keep skin on). Keeping the skin on with these softer fruits stops them from breaking apart when you cook them.

4. Mix spices, crushing the seeds in the mortar and pestle and stirring in nutmeg and ginger. Sauté spices in butter/ghee/coconut oil at a medium heat until fragrant. Add fruit (limes and bananas face down). While the fruit is sautéing, take a frying pan and gently toast the coconut flakes.
5. Place a few pieces of fruit on everyone's plate, squeeze the toasted lime all over the fruit and sprinkle with coconut (and the physalis or star fruit on the side if you like!)
6. Serve with crème fraiche, sugar-free yogurt, or whipped cream.

Drinks

DIGESTION TEA

Stimulate your digestion with these digestive spices.

Ingredients

- 1 tsp. coriander seeds
- 1 tsp. cumin seeds
- 1 tsp. fennel seeds
- Optional: ginger root
- 3 cups of water

Method

Simmer in a saucepan for 5 minutes, drain and sip—between meals.

KOMBUCHA

This is a delicious fermented tea drink, which is fizzy and scrumptious. It is great for the gut biome and is SO expensive to buy, that I started making it. It's really fun to play with the flavors. You can do chili and ginger, pureed pears, grape juice, etc. Below is my favorite blend, which I have found is also kid-friendly!

Ingredients

First Ferment

- 3½ quarts water
- ½ quart plain kombucha from last batch
- 8 bags of tea (black or green)
- 1 c. sugar
- 1 scoby (kombucha culture)

Second Ferment

- Shatter-proof glass bottles (I use the ones that brewing stores sell with flip-top lids)[63]

For each 15 oz. of first-ferment kombucha, you will need:

- 1 tbs. apple sauce
- 1 tsp. sugar (less fizzy option: 2 tbs. apple sauce)
- 1–2 inches cinnamon stick

Method

First Ferment

1. Bring water and sugar to boil. Remove from heat. Add tea bags (if they have dangly bits, chop those off first, otherwise they get all entangled). Remove bags when tea is room temp or cooler.

2. Put this tea and the ½ quart plain kombucha from last batch (or from your kind friend who gives you the scoby) with the scoby in a glass container. Cover it with breathable fabric.

3. Leave for about a week.

Second Ferment

1. Sterilize your shatter-proof bottles.

2. For each 15 oz. of first-ferment kombucha add:

    - 1 tbs. apple sauce

63 Do not use mason jars!

- 1 tsp. sugar (or 2 tbs. apple sauce)

3. 1–2 inches cinnamon stick

4. Leave at room temp for about a week—taste it. The sugar gets "eaten" as it ferments, so turns more sour over time.

5. Refrigerate to stop fermentation.

CHAI TEA AND GOLDEN MILK

Ingredients

Golden Milk[64]

- 1 cup whole milk or non-dairy equivalent
- ½ stick of cinnamon
- 3 gently crushed pods of cardamom, or about 9 seeds
- 4–5 shards of fresh nutmeg, or ¼–½ tsp. of ground nutmeg
- ½ teaspoon of vanilla essence
- ½ teaspoon of turmeric
- 3 black peppercorns or a twist or two of black pepper (makes turmeric more bio-available).

CHAI TEA

- 3 cloves and a black tea bag—add these if you're making tea, but do not use these at bedtime, they will negate the sedative spices

You can use ground nutmeg or cinnamon, but they don't blend with the milk and tend to sink to the bottom of your cup. Using the shards and the cinnamon stick gives a more subtle infused flavor!

Method

GOLDEN MILK

1. Put milk and spices into a saucepan and warm to boiling point (makes cow milk more digestible). Watch with the nut milks, they might separate on boiling, so turn off just as it starts to rumble!

2. Strain into a teacup and enjoy! By the way—if you are

64 Will keep you asleep.

using unsweetened nut milk, you may want to add ½ tsp. of maple syrup or equivalent.

CHAI TEA

1. Put milk, tea bag or 1½ tsp. of tea, and spices into a saucepan and warm to boiling point (makes cow milk more digestible). Watch with the nut milks, they might separate on boiling, so turn off just as it starts to rumble!

2. Strain into a teacup and enjoy! By the way—if you are using unsweetened nut milk, you may want to add ½ tsp. of maple syrup or equivalent.

HEY!

Have fun with this! Connecting with your food is easy—it's fun to explore and engage and learn. Bulk cooking these dishes helps you to get on with your life—make the granola and some soups for your week on a Sunday afternoon. When you remove the choices, by setting yourself up for success, you'll have the time and energy for YOU!

Let me know how it's going:

http://www.facebook.com/TamsinAstor

http://www.twitter.com/TamsinAstor

http://www.instagram.com/TamsinAstor
(where most of my recipes and food posts are)

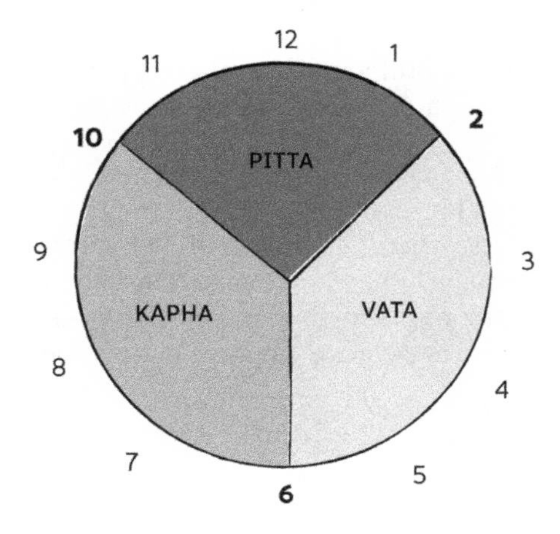

JOURNAL QUESTIONS TO HELP YOU DIVE DEEPER

WHY JOURNAL?

Because you want to develop emotional agility, which refers to your ability to approach your inner experiences mindfully and productively, rather than suppressing them. Approaches that I detail in this book, such as yoga, meditation, cognitive strategies (e.g., challenging your negative thoughts about yourself or your resistance to changing an unhealthy habit or resisting a good habit), and journaling can all contribute to a happier mind!

Journaling can help you problem-solve, clarify your thoughts, or start to see whether the people in your life are positive or negative.

Research shows that the practice of journaling has impressive positive results! From journaling to manage your fears and your traumatic experiences, which has significant health, social and behavioral outcomes (including improving your immune, lung, and liver function), to journaling before you take a test (which improves your scores), to reducing depressive symptoms, to reducing absenteeism from work!

JOURNAL 1.

What is a habit that you want to add to your life? Why? What do you need to unleash your power? How are you going to make it happen? What step are you going to take to make this a reality today?

JOURNAL 2.

What is a habit that you want to stop? Why? What could you replace this habit with? (Your brain is wired for the habit, so replacing it with a better habit is MUCH easier than totally getting rid of the bad habit!)

JOURNAL 3.

Who is your **"ought self"**—the person you think you "ought" to be? How do you look, feel, dress? What do you do? Where do you live? Who is your **"ideal self"**? When are you happiest? What are your habits? What are you doing? What are you doing when you are totally lost in the experience? What's the gap between these two? What can you do to minimize this gap?

JOURNAL 4.

When you have taken on a big challenge (e.g., weight loss, running a marathon, moving to a new house, completing a degree), what were the circumstances that made it successful? Who helped you, did you track your success, were you competing, did you have to do something if you lost the bet?

JOURNAL 5.

In what ways is your environment supporting or cramping your healthy choices? (Do you keep snack food in your house that you have to resist? Is your gym located between home and work? Does your partner pour you a large glass of wine when you walk in the door, which leads to you eating badly and not going to yoga?)

JOURNAL 6.

What does it feel like when you commit to your goals? Why?

JOURNAL 7.

Where in your life do you prioritize your self-care, that is, put on your own oxygen mask first? And last? Why?

JOURNAL 8.

How do you numb your feelings? Eat, sleep, anger or projection onto others, watch Netflix, an addictive habit?

JOURNAL 9.

What situations create the most stress in your life? Can you avoid, decrease, or delegate these situations? What role do you play in elevating the stress in these events?

JOURNAL 10.

When you experience stress (remember that good things—like babies, new houses, new jobs—are also stressful), what do you do?

A. Guide: Perhaps you look back and wish you hadn't done something. Maybe you obsess and get stuck in the current moment. Do you sink into the idea that everything will be okay, when you get into a better relationship, lose that weight again, etc.? Do you reflect and evaluate your role and look for the learning or silver lining?

JOURNAL 11.

Do you struggle with FOMO (fear of missing out)? What are you afraid of missing out on, if you stay up late? In what ways are you productive after 10:00 p.m.?

JOURNAL 12.

Why do you eat if you're not hungry? What's really going on? What are you hungry for, when you eat without hunger?

JOURNAL 13.

What does nourishment mean to you? Explore what it means to be nourished in your body and mind, your other senses.

JOURNAL 14.

When I have a consistent exercise routine, I feel...

JOURNAL 15.

What exercise makes you feel strong, gorgeous, and with a deep connection between your body and mind? (Maybe this was something you did as a kid.) What are the main components that you love? Group, individual sport? Indoor or outdoor? Competitive, music?

JOURNAL 16.

What are five things you are grateful for? (Do this every day! You'll build the gratitude muscle.)

JOURNAL 17.

What person, place, and thing are you grateful for in your life?

JOURNAL 18.

Who are the supporters in your life, and what do they do for you?

JOURNAL 19.

Who are the enablers in your life? What do they do that does not feel supportive (e.g., encourage you to overeat, drink too much, tell you you're fat/stupid/unkind/belittle your efforts, gossip, or look for the negative), and can you create some boundaries or space between you and them?

JOURNAL 20.

When are you in flow? When do you forget about time because you are so engaged in the task? How can you add more of this in your life?

ACKNOWLEDGMENTS

I would like to start by thanking the people who pushed me to write this book—including my lovely editor Lisa Bess Kramer at Cleveland Edits—the voices who kept telling me I had a unique take on life and a great writing style and should do it (I resisted for a while, but eventually got it together!). I would also like to thank Craig James who, like me, knew he had something inside him to share, and, like me, attended an event where we identified this and then both walked up to a notice board and pinned up "write a book" in the "scare your soul" section! I'd like to thank my copy editor Jenna Skwarek, who was instrumental in helping with style formatting and citations. Also, I'd like to thank my friend Jen, who helped with the recipes, and my dog and my coffee pot, who both kept me company when I was writing pre-dawn during the dark, cold Cleveland winter months.

I would also like to offer my profound thanks to Demos Ioannou, who, after getting to know me, my passion and commitment to this work and my writing, connected me with the inimitable Brenda Knight at Mango Publishing, which ultimately led me to a multi-book deal!

And, penultimately, a heartfelt thanks to my beloved parents who are ALWAYS at my back. Most importantly, I'd like to thank my children—M, J, and D—for whom organizing my life and living in a healthy way is why I do this work!

ABOUT THE AUTHOR

Tamsin Astor

Tamsin is an immigrant, a mother, a lover of live music, travel, tattoos, dark chocolate, coffee, and homemade bread. She fought for many years against the notion of creating rituals and routines in her life, because, frankly, it sounded boring and middle-aged. But, having absolutely discovered the power of creating nourishing routines and seeing how they free up her life, she is now the Chief Habit Scientist who advocates the power of rituals to anyone who will listen.

Read more about her here: http://www.TamsinAstor.com

NOTES

NOTES

NOTES

NOTES

NOTES

NOTES

NOTES

CPSIA information can be obtained
at www.ICGtesting.com
Printed in the USA
BVHW030357190819
556114BV00003B/3/P